Wagon-Making in the United States

Dedication

This work is dedicated to Paul A. Kube, to the Gruber family, and to all those who have contributed to the preservation of the Gruber Wagon Works and its history.

In substantive commitment to this dedication, the Memorandum of Agreement toward publication of this work, dated 14 August 2004, stipulates that:

All royalties earned by the Work shall be deposited directly to the savings account of the Berks County Parks & Recreation Board, and are designated specifically and exclusively for the support of the Gruber Wagon Works.

WAGON–MAKING
IN THE UNITED STATES
DURING THE LATE-19TH THROUGH
MID-20TH CENTURIES

———————

A STUDY OF THE
GRUBER WAGON WORKS
AT
MT. PLEASANT, PENNSYLVANIA

BY

PAUL A. KUBE

WITH CONTRIBUTIONS BY

CLAYTON E. RAY AND CATHY L. WEGENER
ASSISTED BY OTHERS

———————

The McDonald & Woodward Publishing Company
Blacksburg, Virginia

The McDonald & Woodward Publishing Company
Blacksburg, Virginia, and Granville, Ohio

Wagon-Making in the United States during the Late-19th through Mid-20th Centuries: A Study of the Gruber Wagon Works at Mt. Pleasant, Pennsylvania

All rights reserved. First printing September 2005
Printed in the United States of America by
McNaughton & Gunn, Inc., Saline, MI

15 14 13 12 11 10 09 08 07 06 05   10 9 8 7 6 5 4 3 2 1

Library of Congress Cataloging-in-Publication Data

Kube, Paul A., 1918–1988
  Wagon-making in the United States during the late-19th through mid-20th centuries : a study of the Gruber Wagon Works at Mt. Pleasant, Pennsylvania / by Paul A. Kube ; with contributions by Clayton E. Ray and Cathy L. Wegener, assisted by others.
      p. cm.
  Includes bibliographical references and index.
  ISBN 0-939923-97-1 (pbk. : alk. paper)
  1. Carriage and wagon making--United States--History--19th century.
  2. Carriage and wagon making--United States--History--20th century.
  3. Gruber Wagon Works--History. I. Ray, Clayton Edward, 1933– II. Wegener, Cathy L., 1959– III. Gruber Wagon Works. IV. Title.
   TS2010.K82 2005
   688.6--dc22

                                                           2005025285

Contents

List of Tables and Figures

Tables and figures for Paul Kube's thesis are
separately numbered as tables 1–5 , figures 1–102,
and Appendix figures 1–4, and are listed on pages
36–39.

Introduction and Context

Introduction

The purpose of this publication is to make widely known and available a unique contribution to the history of wagon-making: Paul Kube's study of the Gruber Wagon Works (hereafter referred to generally as the Wagon Works), prepared in 1968 as his thesis for the degree of Master of Education at Millersville State College (now Millersville University), Millersville, Pennsylvania.

A personal note about the author of this introduction seems in order here, not to put oneself forward, but as a disclaimer to any special qualification other than persistent interest, and as an explanation of the genesis of this project. My part can be traced at least as far back as 1934, when my mother drove Molly and Topsy, a team of dappled gray grade Percherons, hitched to a farm wagon, a mile or so from one home to another in Henry County, Indiana, carrying household belongings and two small children, one of whom was I. From there by great good luck, though not by wagon, I went to Harvard University (Ph.D. in geology), and ultimately (1963) to the Smithsonian Institution (curator in paleontology; for more detail see Eshelman et al., 2002), where it eventually proved possible, with determination, to work in the center of Washington, DC, while living 50 miles and a world away with draft horses and wagons, demonstrating that you really can't take the country out of the "boy." In 1983, putting the cart before the horse, to coin yet another cliché, we bought for $250.00 and hauled from Scotland Neck, North Carolina, to our then home (and storage in a bedroom) in Arlington, Virginia (within walking distance of the Pentagon and National Airport), our first wagon, a nice Hackney, disassembled, the box body on top of our Volkswagen bus, and the axles sticking out of the open sliding side door. The horses, Suffolk mares, followed in 1984, by then to a farm near Falmouth, Virginia. On retirement in 1994, I was able to return even more actively to second childhood.

Having acquired more wagons (not to mention carts, bob sleds, and obsolete implements), at the rate of about one a year, and having spent a career in bookish research, I felt a compulsion to learn more about wagons, and began to browse through the files of the Department of Transportation in the Smithsonian's National Museum of American History (courtesy of Roger White). There I quickly made two discoveries: first, that the literature on American farm wagons is surprisingly meager; second, a copy of Paul Kube's thesis sent in 1968 to then-curator Don Berkebile by Henry Kauffman (Kube's professor; see Appendix I). In juxtaposition to the first, distressing, discovery, the second was literally finding buried treasure. As I gradually learned more, I became convinced that such a gem had to be published — hence this project, leading to acquaintance and collaboration with Cathy Wegener and with many other wonderful wagon enthusiasts.

Although Paul Kube's thesis is a scholarly work that can stand perfectly well on its own without support from us, we have nevertheless felt that some ancillary material, in part stemming from events since 1968 (most importantly the move of the Gruber Wagon Works to its present location), could usefully supplement his work. This material includes a sketch map showing place names important to the story of the Wagon Works (Figure I), a brief historical review of wagon-making, a biographical sketch of Paul Kube (Appendix I), and a list of some Gruber products identified by serial number (Appendix II).

While adding our modest afterthoughts to Kube's thesis, we have been mindful of the risks and responsibilities inherent in presuming to publish posthumously the intellectual property of any author. First, one has to ask whether its publication will be a disservice to the memory of the author, no longer able to speak for himself. Has the work become dated, obsolete, superceded

4

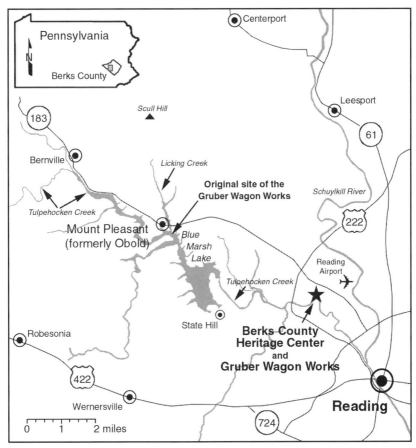

Figure I. A map of part of Berks County showing important place names mentioned in this book. The whole of Berks County is indicated in the inset map of Pennsylvania. The city of Reading lies 55 miles northwest of Philadelphia. Modified from a map provided by Bill Kochan, Director, Mapping Department, County of Berks.

by subsequent work? In this case, besides our conviction that the answer was a resounding "No," there was clear, independent evidence of the profound value of the thesis. For example, it provided much of the foundation for a later Ph.D. project in industrial archaeology (Smith, 1971), and for the supportive documentation justifying the designation of the Gruber Wagon Works as a National Historic Landmark and warranting its move by the US Army Corps of Engineers (John Milner Associates, 1978:11), and has been cited in a recent book on manufacture of horse-drawn vehicles (Kinney, 2004:373).

We feel that Kube's clear, straightforward descriptive prose stands up well with the passage of time, so well that it can take its place honorably alongside the justly celebrated work of Sturt (1923), albeit without the sociological dimension. Further, we have found nothing in print that preempts Kube's work. Although the Grubers obviously were well known and well regarded in their heyday (see, for example, Fegley, 1916), we know of no other detailed and comprehensive description of their work, or that of any other manufacturer.

Sturt of course never regarded himself as a craftsman. His interest was much more in the people and society, and he regarded himself as a writer by profession (see the foreword by Thompson in late editions of Sturt). Neither Sturt, nor any of the writers of wagon books, have concentrated on details of the work. Most have focused on the appearance of the finished product and/or on the business history of the company and personal stories of founders, owners, and executives. Company catalogs and company-sponsored publications, although interesting, generally highlight the virtues of their products, rather than details of how they were made.

The only comparably detailed published descriptions of the details of wagon-making known to us are those of the Foxfire wagons in Wigginton (1973) and Wigginton and Bennett (1986), and those works complement Kube's, as they describe one-man, one-wagon construction much as it would have been for Franklin Henry Gruber before 1883 when he began working in a larger shop. Thus, Kube's work stands alone in its descriptive detail of traditional wagon-making as a small, local industry.

These qualities persuaded us at an early stage that Kube's work should be published intact under his authorship alone, not with us as tardy collaborators. Thus, we have purposively laid the lightest possible editorial hand upon his text and illustrations, making only minimal mandatory changes, including a very few factual corrections, none of which alters the character or intent of Kube's writing.

Examples of such trivial changes, in addition to rare typographical errors, include the substitution of the more common spelling "height" for his "heighth," usage in most cases of "that" in place of "which," insertion of a semicolon or a full stop preceding "however" in order to break up overly long sentences, and changing his descriptions of tools and machinery from past to present tense, as in virtually every instance the thing described as "was" happily still "is."

Equipped with 20/20 hindsight, it would have been easy to find ways to "improve" Kube's text. For example, one could wish for discussion of camber, gather, and dish in wheels, or for more emphasis on hay flats, the Grubers' flagship product, but Kube was not a practicing wagon-maker, and on balance, it is probably providential that he was not — the people who spend their lives doing skilled work have rarely seen fit to write down the details of just how they did things that seemed self-evident to them. Thus, it is best to be thankful for the incomparable contribution that Kube did make, rather than lament this or that seeming omission.

More debatable perhaps, but motivated by the same desire to maintain the coherence of his work, was our decision to retain Kube's original illustrations. These have been largely duplicated and in many cases improved upon by professionals since 1968. However, these photos (more than 200) and diagrams are readily accessible as an independent resource through John Milner Associates (1978), the Library of Congress (under the rubric "American Memory"), and the National Park Service's HABS/HAER program (acronyms for Historic American Buildings Survey/Historic American Engineering Record). We concluded that any improvement resulting from partial or complete substitution of illustrations would be more than counterbalanced by the loss of integrity to Kube's scholarly product.

As with the text, we "corrected" a very few errors in the illustrations, notably Kube's figures 2, 14, 84 (his 85), and 87, which we think were originally printed upside down, figure 11 which we have rotated 90° clockwise, and figures 84 and 85, which were interchanged. As we did not have access to Kube's negatives, all photographs reproduced herein from his thesis were derived from printer's negatives, making it necessary to blur them slightly to minimize the effect of moiré.

Now, grant, If you will after reading it, that Paul Kube did an admirable job on what he did, why is it so important? The answer to that lies in the improbable combination of factors that eventuated in bequeathing to posterity a uniquely rich archive (structures, tools, methods, documentation, and products) not only of wagon-making, but representative also of late nineteenth century manufacturing in general.

First, of course, had to be the founding, development, and continuation to a remarkably late date of the Gruber Wagon Works. A very instructive parallel, and ultimately equally instructive divergence, may be drawn between the Grubers and the Studebakers. The business and human sides of Studebaker history have been repetitiously reported (Cannon and Fox, 1981; Critchlow, 1996; Erskine, 1924; Kinney, 2004; Longstreet, 1952; Wheeling, 1991), though not the details of production à la Kube. Both family businesses were founded in the German roots of southeastern Pennsylvania by single individuals starting small as farmer-blacksmith-wheelwrights, and both remained family businesses through having several sons with the interest, talent, and familial compatibility to build on the fathers' foundations. From there the divergence began, as Studebaker grew very large and diverse, followed by moving early (1920) out of horse-drawn[1] vehicles and aggressively into the automotive age, with considerable success for some decades, followed by complete collapse. The Grubers, by contrast, carried on with little change to the end, never growing significantly larger, never conspicuously modernizing their production facilities, and, although tentatively entering into truck body production (along with automotive wrenches; see Bruce Hunsberger, in press), never completing the transition, continuing wagon-making to 1956 and repairing through 1971. Some concept of their divergence in size can be had from the fact that Studebaker produced 3,300 wagons in 1868 (many more in later years), and its new factory of 1872 had 40 forges (Kinney, 2004:202), whereas Gruber is said to have produced some 100 wagons in its peak year of 1915, and never had more than 4 forges (Kube, this volume). The Studebakers moved with the suddenly and drastically changing trends of the marketplace as did some other wagon-makers, generally proving to be the alternative to extinction. By contrast, the Gruber Wagon Works remained in business much as before for half a century beyond the great changeover. They were able to do that partly through their own conservatism, family commitment, and modest ambitions, and partly through location in a farming region where the horse-drawn era continues strongly even to the present; the Studebakers also were convenient to a major continuing, largely Amish, market, but chose a different pathway.

Second, Paul Kube emerged as precisely the right person at the right time (see Appendix I). He was a mature student and himself a craftsman, pre-adapted to appreciate, understand, and record the facts about the Wagon Works just in time, while it was still operational. We present that key contribution in this book.

Third, the Gruber Wagon Works somehow escaped the common nemesis of nineteenth-century factories — catastrophic fire. Wagon factories were conflagrations in waiting, and many if not most met that fate. The combination of minimal firefighting capabilities, frame structures, large stores of bone-dry lumber, shavings and wood dust, paint and varnish, hot iron in contact with wood, and coal forges in the blacksmith shop was a formula for fire. A casual review shows complete or crippling destruction by fire at Bain in 1892 (Wheeling, 1994A:170); Hackney in 1890, 1921, and 1945 (Daniel, 1979:32, 54, 72); Mandt in 1883 (Homme, 1947:47); Mitchell & Lewis in 1878 (Wheeling, 1997:158); Schuttler in 1850 and 1871 (Wheeling, 1993:155, 156); Studebaker in 1872 and 1874 (Kinney, 2004:202–203); and Thornhill in 1899 and 1910 (Wheeling, 1995:56). Even though

Figure II. The Gruber Wagon Works as it appears today (2 July 2005), viewed from the west. This view is similar to that in Figure VI. Photo courtesy of Joel Palmer and Glenn Riegel.

these and other companies survived and rebuilt, the historical loss was irretrievable, as the rebuilding generally entailed modernization of structure and machinery.

Fourth, the threat of inundation by the proposed Blue Marsh Lake, a reservoir that the US Army Corps of Engineers proposed to create on Tulpehocken Creek, provided the catalyst to energize the wide and intensive reaction that led ultimately to the commitment in time, money, planning, and organization that was essential to the move in 1976 and preservation of the Wagon Works (Figure II). Without that timely motivation to rally support for preservation of the building and its contents, it seems doubtful that such a heroic feat would have been accomplished. Without the alarm caused by the impending inundation, the likely alternative, if not drowning, surely would have been loss sooner or later through deterioration, vandalism, fire, or dispersal.[2]

Finally, through the impetus of the Society for the Preservation of the Gruber Wagon Works, inspired by the foresight of Elsie M. Rhoads Gruber, the book by Carol Hunsberger (2005) complements and supplements our project perfectly by providing a wealth of historical information about the Gruber family, the Wagon Works, the people, and the area. Rather than inserting a plethora of citations throughout our book, it should be understood that Carol Hunsberger (2005) should be consulted at every turn. Further, Charles Cook's video about the move of the Wagon Works building and its contents is an invaluable resource (Cook, 2005).

Thus, in retrospect, it seems little short of a miraculous cosmic coincidence that the many imperative factors did in fact come together to give us the priceless archive, of which Paul Kube's thesis, presented here, is the cornerstone.

---

[1] In my naïve search in numerous dictionaries for authority to choose among "horsedrawn, horse drawn, and horse-drawn" I found no entry in any of the major dictionaries (Oxford, Webster, Funk & Wagnall's, Random House, American Heritage, Century, World, Winston — to name but a few) for this common term amongst the plethora of horse-combinants, some highly abstruse, rare, or bizarre (including horse-milliner); the closest being "horse drawing," apparently a Canadian term for horse pulling contests. I did finally discover "horse-drawn" in two less-renowned dictionaries (Chadsey et al., 1947:294; Kahn, 1990:266). Otherwise, "horse-drawn" was found concealed in some dictionaries, Webster's 3rd and American Heritage for example, used in defining "wagon," thus violating the dictionary-maker's canon never to use an undefined term in defining another (the "Word Not In," or WNI rule; Landau, 2001:160) Among more prestigious dictionaries, Mathews (1951:835–840) managed to use more than five pages on "horse," mostly devoted to combinants, without mentioning horse-drawn. There and elsewhere the list of combinants is almost identical from one to another, strongly suggesting "follow-the-leader" (negative plagiarism?), and missing the commonest combinant of all. The hyphenated form, "horse-drawn" seems to be commonest in the primary literature and thus is adopted here, although all three variants are well represented in recent issues of *Small Farmer's Journal*.

[2] Now seems to be the best opportunity to put on record a remarkable footnote to the Corps of Engineers' project that otherwise is documented in detail elsewhere, most especially in John

Milner Associates (1978). In command of the Philadelphia District, 1968 to 1972, and later at increasingly higher levels in the Corps of Engineers through retirement in 1980 at the rank of major general, was James A. Johnson, whose family had owned and operated the Stoughton Wagon Factory in Stoughton, Wisconsin, where General Johnson grew up. Although this remarkable convergence did not influence decisions by the Corps of Engineers, it could have done no harm that wagon-making did not have to be explained to General Johnson. (As a footnote to this footnote, it should be mentioned that the Stoughton Trailer Company, a major builder of truck semi-trailers today, is an indirect descendant of the Stoughton-Mandt wagon building businesses of Stoughton. Although the rather complex business history of wagon-making in Stoughton is beyond our scope here, the present trailer building business initially occupied the last buildings of the Stoughton Wagon Factory, which, along with their records, were destroyed by fire in 1967.)

Some Notes on Wagons and Wagon-Making

Wagons, in the sense of 4-wheeled conveyances, as opposed to 2-wheeled carts, both drawn by draft animals (or at times by humans), are known from early antiquity by archaeologically preserved remains, mostly funerary, and through ancient artwork. In earliest times wagons seem to have served primarily as status symbols for royalty and religion, not as utilitarian vehicles. This period is beyond our scope here, but an introduction to it can be had through the works of Piggott (1992) and Stratton (1878). Only with the slow increase and distribution of wealth, the introduction of spoked wheels, the invention of moveable front axles, pivoting around a centerpoint to make possible smooth turns, and the development and spread of passable roadways, did wagons become common and widespread for everyday service. Even then, carts and sledges remained the poor man's necessity up to the end of the animal-powered era, because of their greater maneuverability and economy (see Lemon, 1972, and McGaw, 1994, for factual analyses of this topic, specifically related to southeastern Pennsylvania).

Thus the ascendancy of the wagon era of interest to us here is for the most part encompassed within some 200 years, from the mid-1700s through the mid-1900s, with its apogee from no earlier than 1820 through 1920. The beginning of the modern period of abundance of wagons coincided with burgeoning technology that made lumber and iron abundant and cheap, and made parts reproducible by newly invented machinery (see Cooper, 1994, on these topics). Further, the demand created by the change from subsistence farming to market farming with its urban customers served by commercial freight wagons of many types, along with developing roadway networks, made the expansion of wagon manufacturing necessary and affordable. The wagon era ended abruptly with the rampaging development of the automotive era, punctuated by the late boom in production of wagons and related horse-drawn equipment caused by World War I, coupled with its instantaneous bust at the armistice. The postwar crash in demand, followed by general depression of the 1930s, spelled the end for the majority of wagon manufacturers. The biennial census of manufactures (United States Bureau of the Census, 1924, etc.) provides very revealing quantitative data on the state of wagon-making in the United States, and is particularly revealing for the years of the precipitous decline. The tabular statistics in these reports are well worth study. Naturally, as everyone must know who participates in analogous governmental surveys of agriculture today, little significance should be attached to exact numbers, but great significance to the overall patterns revealed. The report for 1921, published 1924, is especially valuable, as it includes retrospective data and analyses, some quotations from which follow:

> The striking decline of this industry since 1909, and especially since 1914, is due in large part to the enormous increase in the use of automobiles and motor trucks, which have supplanted horse-drawn vehicles to a very great extent in cities and towns, and to some extent even in agricultural communities (1924:1008).
>
> . . . .
>
> The number of complete vehicles manufactured has declined from census to census since 1909 (Table 853). This is true not only of the total for all classes but for each class separately, with the single exception that the production of "governmental, municipal, etc.," wagons tripled between 1909 and 1914 [presumably military demand] . . . . although farm wagons decreased in number from 1909 to 1919 much more slowly than carriages and buggies, the decrease from 1919 to 1921

for each class of vehicles was at approximately the same rate — 82.6 per cent for farm wagons and 84.1 per cent for carriages and buggies (1924:1008–1009).

These reports record much significant detail, although not strictly comparable in manner of recording through the years. One relevant example is the decline in the number of establishments that were engaged primarily in the manufacture of complete horse-drawn vehicles of all kinds (Table I).

For Pennsylvania, always among the leaders, the number of establishments reporting was 139 (1921, number one, followed

Table I. Number of Establishments in the United States Engaged Primarily in the Manufacture of Complete Horse-Drawn Vehicles of All Kinds from 1849 through 1937.

YEAR OF RECORD	NUMBER OF MANUFACTURERS
1849	1,822
1859	7,222
1869	11,847
1879	3,841
1889	4,572
1899	6,204
1904	4,956
1909	4,870
1914	4,601
1919	2,286
1921	826
1923	396
1925	152
1927	117
1929	88
1931	70 or 75
1933	53
1935	45
1937	43

Note: From 1921 onward establishments with products under $5,000 in value were not included, thus eliminating many small local makers from the count.

by New York with 128), 8 (1931), 6 (1933), and 5 (1937), whereas the *American Carriage Directory* of 1898 (Volume ll) listed as makers of wagons in Pennsylvania some 1093 establishments (Green, 2005). Most wagon-makers that survived, did so in name only as they scrambled to convert to production of petroleum-powered vehicles, and no longer reported as horse-drawn vehicle establishments. These survivors could not afford nostalgic preserva tion, so their facilities converted with great speed from woodworking to metal-working machinery, and to mass production of motor vehicles or parts and trailers. Those that made it through the bottleneck are so transmogrified that only the name remains to connect them to the horse-drawn era.

There was minor feedback from the automotive industry to wagon-making, for example, in the introduction of auto-steered front axles and pneumatic tires in late model wagons. But generally speaking the relationship was replacement of wagon-making by automotive vehicle-making.

As with all generalizations, however valid, there are exceptions. A few wagon makers persisted in their traditional mode: Thornhill and Huntingburg to the 1950s, for example, though late models had to compromise due to unavailability of parts. Montgomery Ward catalogs continued to offer Huntingburg wagons into the 1950s, and the Wagon Works remained in business through 1971.

The remarkable thing about the Wagon Works is that it remained to its end at that very late date essentially unchanged as a small, minimally-mechanized industry serving a very localized clientele in a few counties of southeastern Pennsylvania.

When the doors finally closed, the shop looked much as it had in the heyday of wagon-making prior to World War I.

Although not strictly comparable to the persistence of the Wagon Works, and not to be found in any governmental survey, there seem always to be a few people and places happily immune to trends and statistics. This applies in many ways to Appalachia, where still after 1970 there were individuals who could singlehandedly create a farm wagon from scratch using minimal tools and materials at hand. This has been documented in detail by the Foxfire students (Wigginton, 1973; Wigginton and Bennett, 1986) and the wagons preserved in the Foxfire Museum. The museum also houses an interesting, much older wagon that has been inadequately studied (Bailey, 1992).

Although as noted the literature on wagon-making in the United States and the preceding colonies is remarkably meager in view of the importance of wagons to the development and underpinning of the nation, this is not to say that it is without significant contributions, some of broad scope, some on closely related topics, as well as some brief but informative studies. Premier among these, and the one book to have if no other, is Berkebile (1978), a mine of authoritative information. The much more limited dictionary by Smith (1988) is useful for British terminology. For specific wagon hardware nomenclature, Spivey (1979) is a convenient resource. Although it includes very little explicitly about farm wagons, the recent book on city wagons of all kinds (Green, 2002) contains much of great value, as many makers produced both. Farther afield, the charming books of the late Eric Sloane on early American life including woodworking, tools, and related topics, are well known and readily available in the economical editions of Dover Publications (see for example their catalog of Winter 2005, page 3). Other major works on woodworking tools include the classic by Mercer (1975, first published 1929), the wonderful dictionary by Salaman (1990), plus Sellens (2002), and the recent book specifically on the tools and machinery of horse-drawn conveyances by Cope (2004).

As for literature specifically devoted to classic wood-wheeled wagons in America, the list is very short. Best covered is the celebrated Conestoga wagon of southeastern Pennsylvania, with its heyday perhaps roughly from 1750 to 1850 (Berkebile, 1959; Omwake, 1930; Reist, 1975; Shumway and Frey, 1968). The wagons were drawn by the distinctive Conestoga horse, never quite established as a breed, and vanished as such (Strohm, 1863; the horse pictured in Plate XXIV was owned by John or Calvin Eshelman, a relative of Ralph Eshelman of Eshelman et al., 2002).

The distinctive Conestoga wagon technology migrated south-
ward at least as far as Virginia with slight modification (Vineyard, 1993). Fascination with this significant early period of
wagon-making even gave rise to the short-lived American
Waggon[1] Association (Wheeling, 1994B). Although not declared
dead, the moribund state of the organization reflects lack of the
critical mass of actively interested individuals necessary to keep
any organization alive.

A good sense of the sweep of wheeled transport in the history of North America can be had from Dunbar (1937), Florin
(1970), and Stratton (1878).

The history of wagon-making (more nicely, wainwrighting)
in the British Isles (essentially England and Wales, as Ireland
and Scotland were limited to cart-making) has been rather better served than that of the United States, in the form of several
more or less comprehensive books, including, after the classic
by Sturt (1923), Arnold (1969), Bailey (1975), Jenkins (1972), Parry
(1979), Seymour (1990), Thompson (1980, 1983), and Vince
(1975, 1987). British wagon-making contrasted conspicuously
with that of the United States in that it was highly localized
in color, style, and detail of construction, remaining persis-
tently in small, little-mechanized shops of traditional craftsmen. Noteworthy for our purposes here is that the great variety of comparatively heavy and complex construction did not
include the simple box bodies that dominated farm wagon
production in the United States, nor did it include hay ladders or hay flats in any way comparable to those of southeastern Pennsylvania. The closest approach to the hay flats appears to be the extraneous barge-bodied wagons (the distinc-
tion between barge and boat wagons seems imprecise) said to
have been introduced in the 1890s from continental Europe,
and factory-produced (Arnold, 1969:8, 22, 23; Berkebile,
1978:296, 297; Jenkins, 1972:16, 17; Parry, 1979:40; Seymour,
1990:98). Vince (1975:128–139), however, presents evidence of
English factory-built barge wagons as early as the 1850s. In
any case, they were not traditional endemic English wagons.

The application of the term "ladder" in wagons is somewhat varied. In England, the term generally applies to the sometimes elaborate, often nearly horizontal extensions front and/or rear, added to carts and wagons for increasing carrying capacity, especially for hay (Arnold, 1969, Plate 7; Bailey, 1975, figures on pages 3 and 14; Bristol Wagon & Carriage Works, 1994, figures on pages 7–20; Jenkins, 1972:103, 104; figures 32, 34, 37, 38, 40, 49, 51, 54, 55, 56; Thompson, 1980, figures on pages 10,11, 27, 28, 34, 35, 40, 49, 51, 52, 53, 54, 60; Vince, 1975, figures on pages 43, 87, 88, 90, 102, 104, 110, 116, 126, 139). In southeastern Pennsylvania the term "ladder" may be applied to the entire hay-body, including its specialized bolsters, its front and rear hinged frames or racks (also sometimes themselves called ladders), and its open outwardly sloping sides consisting of widely spaced spindles or slats mortised into longitudinal top and bottom rails, each side reminiscent of a ladder in appearance (Figure III). These sides are tied across front and rear by transverse top rails, against which the simple front and rear racks lean when in use. A long springy hay pole may be placed atop the loaded hay, inserted at suitable height in the front rack, bent downward to compress and stabilize the load, and secured on the rear rack. This kind of hay-body is described and illustrated by Long (1972:375–376) and discussed by Shank (1972:23), who applies the name galluse(s) to the rack(s). In demonstration of the endless variety of wagon nomenclature, I have found this term as such nowhere else. Most dictionaries have it as a variant or derivative of gallows, and only spottily in its to-me familiar application to pants suspenders, especially those on bib overalls. These hay ladders are capable of being mounted on any common farm running gear, and are provided with a narrow plank floor. As a personal aside I can testify to one detrimental feature in this design — surely I am not the only one so inept as to have stepped more than once inadvertently through the hay into thin air outside the narrow floor and between spindles.

These Pennsylvania hay ladders would appear to have no historical connection with the eighteenth-century English open

Figure III. A traditional Pennsylvania German hay ladder in front (A), back (B), and right side (C) views. The racks rise some six feet above the homemade floor, and the top rail of the horizontal ladder is 16 feet long. The rear rack (B) is missing its top cross member. The ladder is mounted on a common farm running gear, in this case a Thornhill. Property of Clayton Ray. Photos by Cyndi Nasta.

spindle-sided road wagons as depicted in Vince (1975, figure on page 20) and discussed by Jenkins (1972:9–12). They are, however, essentially identical to a persistent, widespread design of continental Europe (notably the low countries, France, and central Europe), from very early times, as seen for example in the early seventeenth-century painting "Moliners" by Jan Brueghel, also Kugler (1996, figure on page 46) for 1843, to late dates (Jenkins, 1972:56, figure 11(b); Seymour, 1990, righthand figure on page 100, obviously copied from Jenkins). Although at least one of these hay ladders made its way west to be used in the Willamette Valley, Oregon (Florin, 1970:34, figure on page 35), they were essentially limited to southeastern Pennsylvania, to which I believe their design and construction were imported unchanged directly from Germany with the Pennsylvania German immigrant craftsmen.

The most luxurious of hay wagons of course was the "hay flat" ("hay bed" in Kube's thesis) of southeastern Pennsylvania, consisting of a sturdy barge or boat body, with fold-down racks, frames, or galluses front and back, very similar to those of the simpler hay ladders. These were the most distinctive product of the Wagon Works (see our cover and figures IV and V), although similar hay flats were built by Reber at nearby Centreport (now Centerport) (Reber, 1992) and by the Columbia Wagon Company, also nearby, in Columbia, Lancaster County (Columbia Wagons, 1917:30, including figure). They undoubtedly were made by a few others in the same area, including Schaeffer, Merkle & Co., of Fleetwood, but not over so long a period and not so well documented. These hay flats have been illustrated also, for example, in Berkebile (1978:167), Hutchins (2004:169, 170), and Wendel (2004:435, upper right figure). They are sometimes said to have been used in Lancaster and Lebanon counties and to have been made in 16- and 18-foot lengths, but of course were made and certainly used also in Berks County, and were made also in 14-foot length at least by Gruber (whose wagons are said to have been used also in neighboring Schuylkill and Lehigh counties). Probably the most widely circulated illustrations of these

Figure IV. A Gruber 18-foot hay flat number 73 mounted on running gear number 46, at the Berks County Heritage Center; front (A), back (B), and right side (C) views. The racks rise some six feet above the floor. Photos courtesy of Joel Palmer and Glenn Riegel.

Figure V. Members of the Madeira family loading a Gruber hay flat with sheaves, probably wheat, in 1920 on their farm at Hyde Park, near Reading, Berks County. Photo courtesy of Jonathan L. Shalter, III.

southeastern Pennsylvania hay flats are the drawings for a children's coloring book by A. G. Smith representing work in the 1880s on the Benjamin and Catherine Firestone Farm, Columbiana County, Ohio, now preserved at Greenfield Village, Dearborn, Michigan (Cousins, 1990: cover, and figures on pages 20 and 25). The hay flat pictured must have originated in Pennsylvania, as did the Firestones.

Although the "hay flat" clearly resembles a European barge or boat body, and not a British traditional wagon body of any type known to me, I am not versed well enough in European wagon construction to know if similar bodies were made in Germany, for example, in the nineteenth century, as I suspect. My guess is that the Grubers brought this design with them from the Palatinate in 1732, but that idea remains to be tested in the future, probably by others with greater knowledge and access to European evidence.

We do know that the Gruber family included several blacksmith-wheelwright-vehicle makers before Franklin H. Gruber, notably John Adam Gruber (1735–1807), Peter Gruber (see Ellis and Evans, 1883:610), and Isaac Gruber (1833–1906).

A few words should be said here about yet another common solution in the United States to the problem of carrying loose hay (and of course other bulky crops including sheaves of cereal grains and bundles of corn stalks), one that is still encountered in horse-drawn farming communities, including those of southeastern Pennsylvania, since the demise of hay ladders and hay flats. This consists of a long, wide (overhanging all wheels), absolutely planar flat bed made of boards either transversely laid across two heavy (generally 3x8-inch or larger) longitudinal sills fixed to the bolsters of any suitable running gear, or longitudinally laid on transoms fixed to the sills. The sills, with or without transoms, serve also to raise the bed high enough to allow full-locking turns, creating what would be termed a "trolley" in England. These beds are equipped with fixed, vertical racks front and back, generally with several cross pieces (often including a wider one at bottom), attached to two widely spaced uprights set into heavy stake pockets. Single wide vertical side-boards may be set along each edge of the bed to accommodate hauling of smaller materials more suited to a box body. These were the familiar farm wagon bodies of my 1930s childhood in Indiana, and I believe elsewhere over a wide area of the United States. These wagons and Hoosier farming in general of that era have been beautifully recorded in the photos of J. C. Allen, many published in the appointment calendars of Mischka Press (for example that for 2004, photos opposite August 9–15 and August 30–September 5), and in book form (Budd and Brock, 1996: especially pages 79, 80, 96, 98, 104, 115, 138, 139, 147). Their construction was well within the capability of most any farm shop (plans for just such a home-built hay rack are offered by Miller, 2000:203–204), although more refined versions were offered also by some of the large purveyors of agricultural implements, and are still produced by small makers of horse-drawn equipment (often on steel wheels). Their successors are the latter-day forage/bale wagons, generally mounted on steel gears with pneumatic tires and auto-steering (though generally still not tracking well at road speed, as anyone knows who has ever followed one).

Aside from the famed Conestoga wagon, the history of wagon-making in the United States is not well documented but seems to be fairly simple in its generalities. The plain box body so familiar in farm wagons of the US became established very early, certainly by 1800, as evidenced by pictures, for example by Latrobe in 1797 (reproduced in Carter et al., 1985:129; and Egerton, 1993:64–65), and by Parkyns in 1795 (reproduced in Posey, 1989:47). These simple boxes could be home-built by anyone with average skill and hand tools and with access to plain planks and minimal ironwork (I have done it). Of course better quality could be achieved by craftsmen in blacksmith/woodworking shops, and later in factories, small and large, especially when finished, uniform dimensional lumber and standardized hardware became widely available and cheap from about 1850 onward. The basic box, however, remained little changed, in harmony with the common pattern in US manufacturing — one style suits all. This style was perhaps not ideal for any one purpose, but was more or less satisfactory for most, analogous to the continuing function of the pickup truck of the automotive age. It could be used for hauling everything from grain to produce, from freight and household goods to people. Buggies came in late and slow, not only because of the direct cash outlay required to obtain a second vehicle, but also for a second, lighter, driving horse.

These wagons, including the more demanding construction of running gear, were produced locally in hundreds, even thousands, of small shops, in many cases as only a semi-occasional part of general smithing and woodworking. Some of these small shops persisted right to the end of the wagon era. However, after the Civil War, the industry became more and more professionalized and specialized, as a result of the availability of interchangeable parts and the ability to reach wider markets through greatly improved shipping possibilities, especially by rail. An impression of the large number of manufacturers can be gained from various sources. For example, at least 42 wagon-makers can be identified in Pennsylvania as late as 1919 (Pennsylvania, 1920), and hundreds of makers are listed for the country as a whole by Hutchins (2004) and Wendel (2004:458–461), both admittedly highly incomplete.

The early and wide establishment of the simple stereotyped box body made it difficult for any maker to distinguish himself, other than through claims of better materials and better workmanship. These attributes were very difficult to put across persuasively enough to be reflected in sales, especially after development of the great mail order houses and implement companies. Thus the room for small local makers diminished radically well before 1900. The Grubers and a few others had a product in the hay flat that was not duplicated by the nationwide superpowers.

Long before the automotively induced collapse of the industry in the 1920s, the industry came to be dominated by fewer, larger producers, a trend that peaked in the years just before and after 1900 (see also the information from the US census reports presented above). Increasing mechanization, followed by westward expansion, rising farm productivity, greatly improved transportation network (notably railroads), and general prosperity created opportunity for growth and concentrated wagon-making among fewer producers serving larger areas. Although none of the great farm implement companies that developed in the latter half of the nineteenth century began as wagon-makers, all added wagons to their line, by acquiring existing high quality wagon-making operations as part of the overpowering drive to offer a complete line of farm equipment; see for example, Hughes (1995) for this aspect of John Deere history, Wendel (1981:30, 400–406) for International Harvester, and Wendel (1991:11, 19, 30, 325) for J. I. Case. Between these companies and the great mail order houses (Sears, Wards, and others), wagon-making came to be concentrated more and more in the hands of a few purveyors, supplied by large factories, either owned by the conglomerates or more or less anonymously under contract (as Thornhill, Huntingburg, and others, with Montgomery Ward;

see their farm catalogs of the 1940s and 1950s). Sears, Roebuck & Company ventured briefly into its own production early in the twentieth century, but for most of its wagon-selling years depended on contracting with various existing establishments, including Thornhill.

Although the definitive history of wagon-making in the US has still to be written, several very useful studies have been done, of limited scope and/or length, or devoted only in part to wagons. Among these should be noted Hafer (1972), Hughes (1995), Kinney (2004), and the several short articles in *The Carriage Journal*, most by Wheeling (1991–2000), plus Anonymous (1974), Frizzell and Frizzell (1977), and Miles (1970). The very recent gratifying wave of publishing includes such notable examples as Cope (2004), Hutchins (2004), Kinney (2004), Sellens (2002), and Wendel (2004). Wheelwrighting has been particularly well served through modern books, including Bailey (1975), Cope (2004), Dewitt (1984), Hendrikson (1997), Jenkins (1972), Morrison and Morrison (2003), Peloubet (1996), Salaman (1990), Seymour (1990), and Thompson (1983).

Even with this flurry of books, one could wish for more information. For example, I have not found adequate coverage of the widely used Archibald wheel, with several patents from 1870 to 1872; the best account known to me is that of Farrow (1895:61–64). One important piece of machinery, introduced in 1891 and with patented improvements through 1916, but not well covered in the modern literature of wheelwrighting (best is Hendrikson, 1997:56–66; a valuable article from 1895 has been reprinted, Anonymous, 1972) is the West cold tire setter, which made possible speedy, precise, and accurate tire setting. Tires could even be reset on existing wheels "while you wait." These machines were widely adopted and the Grubers acquired a used one in 1908. Some are preserved in museum collections, for example in the Ringling Museum of the Circus, Sarasota, Florida, and a few are even still, or again, in use by the resurgent modern wagon-making trade. Those known by me to have them in use are:

Cornett Buggy Shop, Cartersville, Georgia; four machines in operable condition, all with original leather diaphragms (Don Cornett, personal communication).

Hansen Wheel & Wagon Shop, Letcher, South Dakota; two of these machines in use (Doug Hansen, personal communication; see also Wheeling, 2000, Telleen, 2001).

Harlin Olson, Bozeman, Montana; machine recently disassembled, cleaned, and reconditioned, with original leather parts intact (Harlin Olson, personal communication).

Valley Carriage Works, at Dollywood, Pigeon Forge, Tennessee; machine recently rebuilt, leather diaphragms replaced with synthetics (Bill Burgess, personal communication).

William Proctor's Coach Manufactory, the Sovereign Hill Museums Association, Ballarat, Victoria, Australia (see Hendrikson, 1997:56–66).

Witmer Coach Shop, New Holland, Pennsylvania; machine extensively rebuilt (Harlin Olson, personal communication).

Although few if any of these enterprises are direct lineal descendants of traditional wagon-making shops, each strives as far as practicable to produce authentic products, as do several other makers.

With regard to research and publication there is promise for the future, as Hutchins (2004) is but the first installment of a series, and Peloubet (1996) is but one contribution in the active program of the Carriage Museum of America in publishing on horse-drawn conveyances and their parts. The *Small Farmer's Journal* and the several books by its founder/editor, Lynn R. Miller (see for example, Miller, 2000), is yet another important publication program serving a vigorous, hands-on horse-drawn community. Additionally, the increasing recognition among historians that technology is a prime engine in driving history has resulted in a growing body of important literature (e.g., Cooper, 1994; Lemon, 1972; McGaw, 1994). Surely, the unprecedented flow of publication, much of it in the last decade, is not merely a flash in the pan, but a harbinger of more to come.

---

[1] Here may be the place to deal with the double "g" spelling "waggon," still sometimes used in the United Kingdom and resurrected briefly in the United States with the founding of the American Waggon Association, after having been consigned long since to colonial history. Noah Webster, as a part of his concerted and continuous efforts to simplify, standardize, and Americanize English, had already focused explicitly on "wagon" as early as 1786 (Unger, 1998:105), and it appeared as such in all of his publications and their derivatives. Although widely accepted in the United States (Pei, 1970: vi), the issue was not closed even as late as the 1890s, when Isaac Funk's Advisory Committee of more than 50 scholars drawn from throughout the English-speaking world included 10 members who voted for "waggon" (Mathews, 1933:93). The double "g" in modern American writing may well be regarded merely as yet another lingering anglophilic affectation.

The Paul A. Kube Thesis

A Study of the
Gruber Wagon Works
at
Mt. Pleasant, Pennsylvania

A Thesis
Presented to
the Faculty of the Graduate Division
Millersville State College

In Partial Fulfillment
of the Requirements for the Degree
Master of Education

by
Paul A. Kube
May 1968

TABLE OF CONTENTS

~

LIST OF TABLES

LIST OF FIGURES

INTRODUCTION

Our contemporary industry is the outgrowth of the techniques, innovations, and attitudes of the early American craftsman. His skills, tools, and shops were the factors of its conception. Many of these early craftsmen and their skills have faded into the past with little or nothing recorded of their passing except what remains of their tools and products. In later years, when these tools and products are recovered, the discoverer will only be able to guess at their origin and development. The American wheelwright is a case in point.

The craftsman who performed in the capacity of the wheelwright faded from the industrial scene with the passing of the horse and wagon era. In the 19th century Berks County fostered many shops that produced carriages, carts, and wagons. Today only one remains. This study is an attempt to preserve the knowledge of the craftsmanship of the American wheelwright, as represented by that one: the Gruber Wagon Works, at Mt. Pleasant, Berks County, Pennsylvania.

The intent of this study was to investigate the development and growth of the Gruber Wagon Works, through an examination of the tools, techniques, materials, products, and people that created it. At the same time an attempt was made to reveal this industry as an excellent example of that stage in the development of industry representing the transition between the shop of the single craftsman and the complex factory.

CHAPTER I

THE WAGON MAKING TRADE

The problem of transporting his goods has been a constant problem of man. Evidence indicates that as early as 5000 B.C., man used a sledge to solve this problem. This solution apparently was satisfactory until about 3500 B.C. at which time man adapted the sledge frame to wheels and the two wheel cart emerged. Within the next thousand years in many parts of the world many attempts were made to construct functional four wheel vehicles. However, they obviously did not lend themselves to the needs of man for evidence reveals these wagons did not find much use except for religious or royal functions.[1]

By the 14th century, trade and transportation increased and the use of four wheel vehicles known as road wagons became prominent in Europe; nevertheless, the ubiquitous two wheel cart continued to be the common means of transportation for the farmer. It was not until the advent of land reforms that the farm wagon came into prominence. In England, this came about in the middle of the 18th century. As Jenkins points out, there were over two hundred Enclosure Acts passed at that time, and they had the effect of increasing the farm crops to the point that the farmers needed more and larger wagons to handle their crops, whereupon the village craftsmen began to produce four wheel farm wagons.[2]

These craftsmen who built wagons were known as wheelwrights. Although there is little evidence to support it, the wheelwright probably emerged as such during the close of the 12th

---

[1] V. Gordon Childe, *Rotary Motion* (Vol. I of *A History of Technology*), ed. Charles Singer, E. J. Holmyard, and A. R. Hall. 3 Vols.; London: Oxford University Press, 1954. p. 206.

[2] J. Geraint Jenkins, *The English Farm Wagon*. Lingfield, Surrey, England: The Oakwood Press for the University of Reading. 1961. pp. 10–11.

century, when the craft guilds came into prominence, the apprentice system began to flourish, and the wheelwrights formed a guild. At this time little iron was used in the making of the wagon, and the wheelwright was the sole craftsman involved in its construction. Later, as iron became more prominent in the manufacture of the wagon, the services of the blacksmith were called upon. However, as late as the beginning of the 19th century the blacksmith had not yet merged with the wheelwright trade in England. According to Sturt his father's shop did not employ a blacksmith at that time. In fact it was not until the middle of the century that the Sturt shop engaged a full-time blacksmith in the building of carts and wagons.[3] It was at this time that the use of iron in the construction of the wagon became so prominent that the blacksmith became an integral part of the wagon making trade. Other trades such as sawyers and painters were used by some shops, although the sawyer soon passed from the scene when the wheelwright began to acquire lumber from mills that cut logs into planks and boards.[4]

These, then were the crafts that constituted the wagon making trade at the time Franklin Henry Gruber apprenticed to his cousin, John Henry, in his shop at Corner Church near Robesonia, Pennsylvania, to learn the trade.

---

[3]George Sturt, *The Wheelwright's Shop.* Cambridge, England: The University Press. 1934. p. 7.

[4]*Ibid.,* p. 40.

CHAPTER II

THE TOOLS OF THE WAGON MAKING TRADE

The wheelwright worked primarily with wood; therefore, with few exceptions he used the tools of the wood working trade plus a few tools he developed to perform special functions. The blacksmith who worked with him needed only the standard tools of his craft. These tools changed little throughout the period of time that the wheelwright craft existed; therefore, they were the tools the Grubers used in practicing their trade.

Wheelwright Tools

The Gruber Wagon Works was well equipped with the hand tools of the wheelwright. They were stored or hung within easy reach of the worker. The space beneath the benches, the drawers in the benches, and nails and pegs on the walls and ceiling about the benches all were full of tools. Beneath the bench larger tools, such as spoke dogs, hub jigs, and dowel cutters, were stored. The drawers of the benches were neatly filled with smaller tools, such as plane blades, try squares, drill bits, wood chisels, and other various tools. The bare unpainted walls were full of pegs, nails, racks, and shelves, which contained chisels, scratch awls, drill braces, drill bits, draw knives, spoke shaves, hammers, mallets, saws, hub boring tools, and various types of patterns as illustrated by Figure. 1.

OLDER TOOLS OF THE WHEELWRIGHT

In addition to the general tools of their trade, the Grubers also used several older tools of the wheelwright. Such tools as

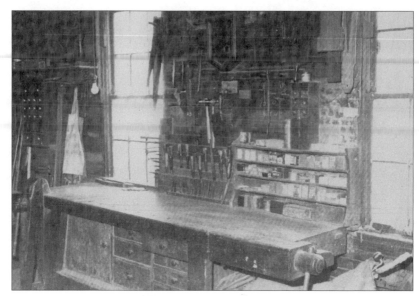

Figure 1. Hand tools of the wheelwright hung on the walls behind his bench.

the bow saw, broad axe and frow were often used in the early shop before the introduction of machinery.

The Bow Saw

If the early wheelwright wanted to cut a curved or irregular shape into a piece of wood, he used the bow saw. This tool was called by many names. It was known as a Holtzapffel, frame saw, turning saw or sweep saw, each name referring to the same saw. Mercer considered it one of the most interesting and important tools of the ancient wood worker.[5] Figure 2 illustrates how the blade was attached. The half-inch-wide blade was attached to the wooden frame by pins, and tension was held by twisting a loop of twine at the other end of the frame. A stick through the twine caught on the center bar held this tension.

---

[5]Henry C. Mercer, *Ancient Carpenter's Tools*. The Bucks County Historical Society. Doylestown: 1960. p. 149.

Figure 2. A typical bow saw used to cut curved shapes in wood.

Axes

Before the advent of the planing machine the axe was used to smooth the surface of large timbers and beams. Since the under-carriage of the wagon had such heavy members, this tool was used by the wheelwright to work on these surfaces. This job required the chisel-edged broad axe with a bent handle as illustrated in the center of Figure 3. The cutting edge was basilled or chamfer-sharpened on one side only so that it could hew to a straight line, and the handle was bent to the side that it was basilled so the person who used it would not get his fingers pinched against the side of the beam.

The process started by scoring the log with the felling axe such as the one at the bottom of Figure 3. This was done by making cuts straight into the log from the side toward a chalk line that was snapped on the log. After the log was rough-squared in this manner the workman placed the log between him and the blade of the broad axe and hewed the surface to a square smooth finish. The smaller axe at the top of Figure 3 is similar and could be called a carpenter's hewing hatchet. It is similar in size to

Figure 3. Axes used to square heavy beams. The axe at the top is similar to a carpenter's hewing axe. The one in the middle is a broad axe, and the bottom one is a felling axe.

those described by Mercer as such; however this tool has a bent handle, probably for the same purpose as the broad axe, unlike those referred to by Mercer.[6] This smaller axe was a one handed axe, whereas, the broad axe was considered a two handed tool.

The Frow

The frow, although not unique to the wheelwright trade, was found to be very useful by the Grubers in their shop. They used it to split logs to make spokes. Mercer identifies it as an "ancient European instrument" and he describes it in detail:

> It is a thick back, ridged dull-bladed steel knife, about fifteen inches long and three and a half inches wide, hafted at right angles upward from its blade, with which by wriggling the short handle to maintain the thickness of the split and clubbing the projecting knife with the Frow Club, shingles, lathes, barrel staves, and short, four to six feet long, clap boards were split (riven) from squared or quartered logs.[7]

---

[6] *Ibid.*, pp. 80–87.

[7] *Ibid.*, pp. 11–12.

The frows used in the Gruber shop were practically identical to those described by Mercer. The blades appeared to be hand forged of two pieces of iron. The eye into which the handle was inserted was formed of soft iron, while the blade was made of steel. The hickory handles were ten to fifteen inches long. Figure 4 shows two of these frows.

These older tools were not used for a very long period of time due to the fact that machines had been designed to do the job that they performed. As soon as the Grubers acquired machinery they retired these tools to the loft, where they remain in storage.

SPECIAL TOOLS

While most of the wheelwright's work was performed with the standard tools of the woodworking or blacksmith trade, some of the features of construction of the wagon demanded the use of unique tools. In some cases these tools were borrowed from other trades and adapted to the wheelwright's use, while others were specifically designed by the wheelwright. Tools in this category in the Gruber shop were the wheelwright's reamer, hub boring tool, spoke dog, bridle stick, hub bench jig, and bench vise.

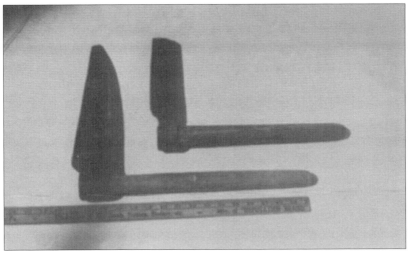

Figure 4. Frows that were used to split logs into pieces of wood that were made into spokes.

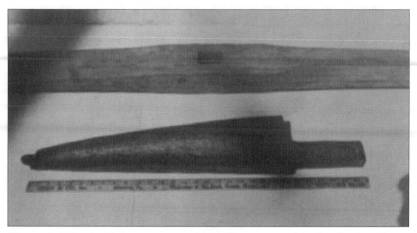

Figure 5. Wheelwright reamer and handle once used in the Gruber shop.

The Wheelwright's Reamer

Although it was used very little in the Gruber shop, the wheelwright's reamer, because of its size, was probably the most outstanding tool in the shop. This tool, hand forged of iron, was thirty-six inches long. It had the shape of a cone cut in half lengthwise. One side of this cone was worked to a fine cutting edge while the other side was left dull. A large rectangular tang was formed on the large end of the cone while a hook was placed on the smaller end. This tool was five and one half inches in diameter at its larger end, and it tapered to two inches at the smaller end. The tang was six and one half inches long, two inches wide, and one and one half inches thick with a three-eighths-inch-diameter hole through the flat side near the end of the tang. The heavy oak handle of this reamer was seven feet long with a rectangular hole at mid-length to accommodate the tang of the reamer. The hole in the handle shown in Figure 5 shows how the hole is reinforced by two dowels driven through the handle near the hole.

Mercer indicates that the English wheelwright used a reamer similar to this one; however, as machines came into use, and the wheel bearings became smaller, this tool became obsolete and fell into disuse.[8]

---

[8] *Ibid.,* pp. 193–194.

A later version of the hub reamer was tried by the Grubers, but it was considered unsatisfactory. This tool was a combination drill and reamer. It was small in comparison to the older wheelwright reamer. It was only fifteen inches long and four inches in diameter at its large end. The small end of the reamer was preceded by an auger drill bit one and one half inches in diameter. This combination tool was designed to drill the pilot hole and ream it in one operation, however, because of its inaccuracy, the Grubers did not use it. The reamer is shown in Figure 6.

Hub Boring Tool

Another special tool used by the Grubers was a hub boring tool. This tool was used to drill the first hole through the wheel hub. It was also used to redrill the hole after the hub had dried and the hole needed to be recentered. The face plate of the jig was placed firmly against the end of the hub. The jig was then clamped firmly in place by the three jawed chuck, which centered the boring tool on the hub. A cutting tool was held in the end of the boring bar, and as it was rotated it cut a true hole in

Figure 6. Later version of a hub reamer.

Figure 7. Hub boring tool mounted on a hub.

the exact center of the hub. The boring bar was rotated by a twenty-four-inch handle as shown in Figure 7.

No machine process ever replaced this boring tool in the Gruber shop, for although it was a hand job, the accuracy of this tool was considered essential for a good hub.

Spoke Dog

The assembling of the wheel demanded that the wheelwright use yet another special tool. This tool was the spoke dog. It was used in the process of putting the felloe on the spokes. It consisted of a hickory handle that was twenty inches long, and an

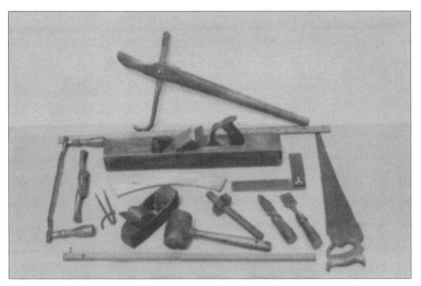

Figure 8. Spoke dog with other hand tools of the wheelwright.

iron bar with holes drilled through at two inch intervals on one end and a two inch hook formed on the other end. The spoke dog is shown in the background of Figure 8.

The spoke dog was used to pry the spokes into line with the holes drilled into the felloe. The hook was caught on one spoke, and the wooden handle was caught over the adjacent spoke; they were then pulled into line with the holes to be fitted into, and the felloe was hammered onto the spokes.

Dowel Cutter

Wheelwrights fastened the felloes of wheels together with dowels since the Roman period of history.[9] The Grubers continued this practice, fastening the felloes together with dowels they made in the shop. Short pieces cut from the end of the spokes when they were trimmed were used for this purpose. They were split into small pieces and driven into the hole of the cutter.

The dowel cutter used by the Grubers was made in the shop. A piece of iron plate approximately three inches by six inches

---

[9]Jenkins, *op. cit.*, p. 25.

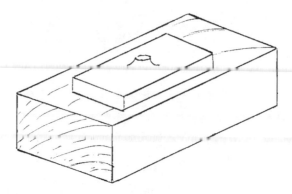

Figure 9. Dowel cutter.

was heated and a punch was driven through the iron. The punch formed a burr as it pushed through the other side of the plate. This burr was filed slightly to sharpen the inside edge of the hole, and the plate was fastened with the burr facing up on a block of wood with a hole drilled through it. The short piece of oak was driven into and through this hole, and as it was driven the dowel was shaped.

The job of making dowels in the Gruber shop fell to the younger men in the shop, using a dowel cutter as illustrated in Figure 9. The dowels made in this manner were stored in boxes on the wall behind the wheelwright benches.

Wedge Cutting Block

Another device that was in constant use in the Gruber shop was the wedge cutting block. Wedges were in constant demand by the wheelwright, and in order to keep him supplied one of the younger men or apprentices was kept busy making wedges. The ends of the spokes projected from the felloe after they were driven into place. These were cut off, and this piece of oak was cut to a length of one and five eighths to one and three quarters inches long. They were split into one-quarter-inch-thick pieces, placed in the notch of the wedge cutting block, and shaped into a wedge through the use of a drawing knife. Figure 10 illustrates such a block as used for this purpose.

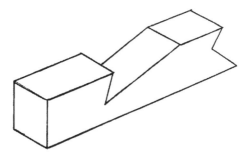

Figure 10. A wedge cutting block. Made from a block of oak, it was used to make wedges from the spoke ends that were cut off.

Spoke Tenoning Tools

Older wheelwrights at one time cut the tongue or outer end of the spoke with a chisel and saw so that it would fit the felloe. Some spokes were fitted with a square tongue.[10] The Grubers found that a round tongue was more satisfactory, and to insure a good fit in the felloe they used a special tool to cut the end of the spoke to fit into the round hole they drilled into the felloe. Figure 11 reveals what they looked like. The spoke tenoning tool was used in a drill brace for a time, then the tangs were modified so that the tools could be used in a drill press. Note the modified tangs in Figure 11.

The tenoning tool was adjustable so that both the depth and the diameter of the tenon could be varied. There were two removable cutting blades: one cut the diameter of the tenon, and the other cut a square shoulder on the tenon. Before the tenon was cut on the spoke the tool was checked on a scrap piece of wood. The diameter was checked by fitting the trial tenon into a hole drilled in another piece of scrap with the drill that would be used to drill the hole in the felloe. If the fit was satisfactory, the wheelwright proceeded to cut the tenons on the spokes. This process was mechanized when the Grubers made their spoke tenoning machine.

---

[10] *Ibid.,* p. 68.

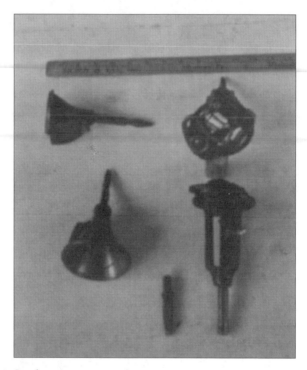

Figure 11. Spoke tenoning tools.

Mortising Chisel

Before the Grubers acquired a mortising machine all square mortises in the hub were cut by hand tools. Among other tools used in this operation was the mortising chisel. This chisel is shown in Figure 8 next to a firming chisel. It was ground to a sharp "V" shape so that it would be able to fit into the square corners of the mortise in the hub and thereby cut them clean. Jenkins refers to a mortising chisel of a different shape that was use by the English wheelwright. His chisel or bruzz was, "a long socket-handled V shaped tool with a blade 5/8 inch wide."[11] Sturt refers to this chisel as a bruzz and describes it as a "three-cornered chisel used for cleaning out the mortices* of a stock and,

---

[11] *Ibid.*, p. 66.

*Author's method of indicating words spelled incorrectly.

to the best of my belief, used for nothing else, unless for enlarging the central hole in the stock."[12] The mortising chisel continued in use after the mortising machine was put into use in the Gruber shop, for it was sometimes necessary to "clean out" a mortise before a spoke would fit into it.

Wheelwright's Bench Vise

Probably the most prominent mark of identification of the wheelwright, aside from his product, is the vise he used. The vise was similar in construction to the carpenter's vise except the vertically mounted jaws projected above the top of the bench eight to ten inches. This variation was undoubtedly due to the difference in the work that was done by the two trades. Whereas the carpenter used the saw and plane to do most of his work, the wheelwright used the spoke shave and drawknife in much of his work. The extended jaws of the wheelwrights's vise permitted the use of these tools with greater ease than the carpenter's vise.

There were four work benches in the bench shop of the Gruber Wagon Works. On the left front corner of each bench was mounted a wheelwright's vise. The vise on John Gruber's bench is shown in Figure 12.

Figure 12 also reveals another difference in the construction of the wheelwright's vise. The turn screw on this vise is mounted high on the jaw of the vise, directly beneath the bench top; whereas, the turn screw on the carpenter's vise is mounted somewhat lower on the leg. Mercer describes the construction and function of this vise:

> It is equipped with a turn screw, which by the right or left twist of a handle, freely sliding in a hole at its enlarged outer end, opens or shuts against the bench leg a heavy, top-rounded plank section forming the jaw or chop block of the vise. By no means as simple as it looks, this turn screw is threaded where it enters the bench leg, but not threaded and turns loose at its neck

---

[12]Sturt, *op. cit.,* p. 102.

Figure 12. The wheelwright's vise.

where it penetrates the chop block, hence, at this latter point the screw is ingeniously "keyed" to the chop block,...by a wooden strip called a garter, round-notched at its inner end, which, penetrating a square mortise in the chop block, engages a groove in the unthreaded screw neck. As a result, a back turn of the screw not only <u>loosen</u> but pulls open the jaw of the vise. It further appears that because an object pinched above the screw would push the chop block outward at its top and hence inward at its bottom, so as to bind the screw, another equipment is necessary. This is a

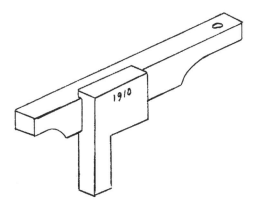

Figure 13. Bench rest used on the wheelwright's bench.

> long, thin, wooden strip, punctured with peg holes,
> extending at right angles from the bottom of the chop
> block. It slides loosely into a slot in the bench leg and
> can be pegged outside the latter to stop the obstructive
> bottom inpush of the chop block.[13]

The strip of wood that Mercer refers to can be plainly seen at the bottom of the vise shown in Figure 12; however, the vise must be disassembled to see how the screw is assembled in the block. The wooden garter referred to by Mercer was found in some vises to be made of metal.

Bench Rest

When the carpenter clamped a long board in his vise he supported the other end of the board on a peg, which he placed in the hole in the right leg of the bench. This helped to hold the board more firmly while it was being worked on. Because the wheelwright's vise held the board above the top of the bench the wheelwright had to devise another means to support the other end of the board.

To do this job the wheelwright devised an "L" shaped bench rest. This rest, shown in Figure 13, was placed in one of the bench stop holes along the length of the bench top.

---

[13]Mercer, *op. cit.*, p. 72.

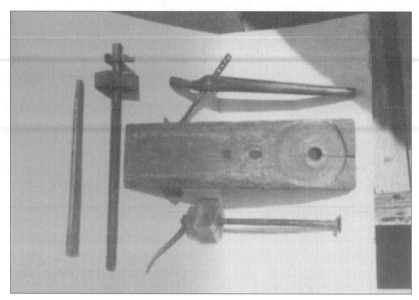

Figure 14. Wheel block or clamping devices used to hold the wheel during assembly of felloes.

The Wheel Block

When working on the wheel the wheelwright had to provide some means other than the vise to hold it. The wheelwrights in the Sturt shop used a wheel block or wheel stool.[14] The Grubers used several devices to hold the wheel depending on what they were doing with the wheel at the time.

When they were assembling the felloes on the spokes they used a block such as shown in Figure 14.

The block was clamped to the work bench as shown in Figure 15. The long bolt with the L nut shown at the bottom of Figure 14, was passed through the center hole in the block and then through a stop hole along the edge of the bench. The wooden blocks shown on the bolt were used as washer-spacers beneath the bench.

The wheel was placed on the block hind side down, and a long wooden rod or iron bolt was passed through the center hole

[14]Sturt, *op. cit.*, p. 104.

Figure 15. The wheel block mounted on the bench and a wheel clamped in place.

of the hub and through the largest hole in the block. The long wedge on the left of the figure was passed through an elongated slot near the lower end of the rod, and the wedge in the slot near the top of the rod was driven in to tighten the wheel so that it could just about be rotated on the block.

There were only two such blocks in the bench shop, so sometimes when it was necessary, the wheel was placed on the floor and clamped down in much the same manner. The hole shown in Figure 16 was used for this purpose. This hole was used mostly for repairing wheels rather than making new ones, for the removal of the old felloes required the use of a heavy maul or sledge which could be swung better in this position.

BLACKSMITH TOOLS

The blacksmith working with the wheelwright did not require any tools other than those normally used in a blacksmith shop. The blacksmith shop in the Gruber Wagon Works was well equipped.

Figure 16. Spot on the floor used to clamp wheels down to work on them.

The shop at one time had four forges, although two were removed when the work of the shop diminished. There were innumerable tongs, hammers, swages, cutters, and chisels hung about the forge as can be seen in Figure 17.

Although there were no special tools used by the blacksmith in the Gruber Wagon Works, there were some that were more interesting by virtue of their use or design.

The Mandrel

The mandrel was a very important tool in the Gruber shop because of its use. It was this tool that gave the final shape to all the hub bands that were made in the shop. The band iron was cut, shaped into a rough ring, welded, and then put on the mandrel where it was formed into a true ring and given the conical shape that was needed to make it fit the tapered shape of the hub.

The Wagon Jack

Although this tool was not exclusively used in the black-smith shop, it was used there more than elsewhere. The blacksmith

Figure 17. Tools of the blacksmith hung about the forge.

was the worker who removed and put on the wheel of the wagon. The tool that was used to lift the wagon so that the wheel could be taken off or put on was the wagon jack. It worked on the simple principle of an eccentric cam action. The basic shape was a forty-five degree angle. In the hypotenuse of the angle a series of steps was cut. The lower corner of the angle had a small wheel attached to it to make it easier to maneuver the jack into position. On the upper corner of the angle the frame was coupled by a cam-like linkage, the vertical leg of which was hinged to the base. The details of the jack are revealed in Figure 18.

The jack was placed in position under the axle of the wagon and placed so that the highest step was under it. The handle was pushed down and the wagon thereby raised clear of the ground. Figure 19 shows the jack in use.

Wheel Stool

One of the difficult and demanding chores that the blacksmith had to do in the wheelwright shop was "ironing the wheel."

Figure 18. Wagon jack used to lift wagons so that the wheel could be removed.

Figure 19. Wagon jack shown in use.

This was done in various ways by different smiths. The process used for a time in the Gruber shop required the use of a unique accessory. The wheel stool was a three-legged triangular-shaped device. The main frame was constructed of six-by-six-inch timbers that met to form an equilateral triangle. Each side of the triangle extended twenty inches beyond the angle so as to extend the surface of the stool to a sixty inch diameter. The top of each leg was surfaced with a strip of iron to reduce wear and burning from the red hot tire as it was placed on the wheel.

The wheel that was ready to receive a tire was placed front side down on this rest. The red hot tire was removed from the fire and quickly set down over the wheel. Before this was done and as it was being done the rest was wet down with water to prevent it from burning. Figure 20 shows a wheel resting on the device as it would in readiness to receive the tire. This device, like so many other tools that were used by the wheelwright and the blacksmith working with him, was no longer used after the process of tiring the wheel was mechanized.

Figure 20. Wheel stool with wheel in place ready to receive a tire.

Machinery eventually replaced most of the hand processes. By the end of the 19th century the Grubers had mechanized most of their processes. They were not the only wheelwright shop to do so. Sturt comments that in 1889 he had mechanized and "from the first day the machines began running, the use of axes and adzes disappeared from the well-known place,"[15] Sturt converted for financial reasons, and the Grubers converted to meet the increased demand for their product, each reflecting the changes in the economy in the individual nations. Whatever the cause, it appears that the machine age had arrived and the wheelwright trade was caught up in it.

---

[15] *Ibid.*, pp. 200–201.

CHAPTER III

SOURCES OF POWER USED IN THE GRUBER WAGON WORKS

By the middle of the 18th century the farmers of Berks County were well aware of the application of water power. A water-powered grist mill was located every two miles along any stream of appreciable size. By 1850, the utilization of water power was made even more efficient through the use of water turbines made of iron.[16] Many farms in Berks County, however, were not located on streams of any proportion, so the farmers had to devise some other source of power. Many farmers in the area of Mt. Pleasant used a horse-powered turnstile. This source of power was applied to the various machinery the farmers made and used.

THE FARM SHOP

The farm on which F. H. Gruber started his wheelwright shop was located near the Tulpehocken Creek; however, it was high above the water level and not close enough to use the waters of the creek as a source of power. The farm had a horse turnstile, and this was the power that F. H. Gruber harnessed in his early shop.

Horse Power

The turnstile on the Gruber farm was located about twenty feet from the barn, in the middle of the barnyard, which was typically located on the lower side of the bank barn of Pennsylvania. The head of the capstan projected about three feet above the ground. It had two five-foot-long poles, spaced one hundred

---

[16]W. Paul Strassman, *Risk and Technological Innovation*. Ithaca: Cornell University Press, 1959, p. 207.

and eighty degrees apart, on the head to which the horses were hitched. Either two or four horses were used, depending on the load that was being placed on the power train. Figure 21 illustrates how the turnstile was constructed.

The capstan shaft extended down into a hole in the ground and at the bottom of the shaft was a gear coupling that attached it to a horizontal shaft leading away from the capstan, in a line parallel to the barn. This shaft was called the jack shaft. It was approximately ten feet long and at the other end was a three-foot-diameter pulley. This pulley was connected by a belt that led upward at an angle along a ditch to another pulley, which was in the barn twenty feet away. A vertical shaft led from this pulley upward to the threshing room floor, where the machinery was located. Here Mr. Gruber placed his table saw, wood lathe, and later the band saw he acquired.

THE NEW SHOP

After he decided to move and enlarge his wheelwright shop, F. H. Gruber investigated various power sources to determine which would be most satisfactory for his use. He finally decided to use the newly developed water turbine. Usher claimed it to be superior to the older water wheel in that it furnished larger amounts of horse power using less water.[17]

The Water Turbine

Mr. Gruber made several inquiries about the possibility of using the water turbine in his new shop. One such inquiry brought the following reply from Mr. Elias B. Schmehl, a millwright located at 145 Penn Street, Reading:

> Reading, Sept. 24, 1883
> Mr. F. H. Gruber
> > Dear Sir:
> > > I saw Mr. Albright this
> morning and he told me that you want a water wheel

---

[17] Abbott Payson Usher, *A History of Mechanical Inventions*. London: Oxford University Press, 1954. p. 382.

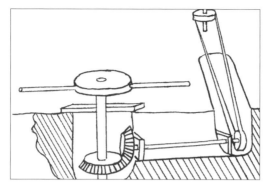

Figure 21. Horse turnstile such as used on the Gruber farm 1870-1884.

to make about 8 horse power. I must see your place
before I can say what it will cost and what sise it takes.
I must see the amount of water that is in the creek.
I will come up there before long and see about it.
Yours truely,
Elias B. Schmehl

Mr. Gruber finally contracted for the delivery of a Reliance
Turbine Water Wheel manufactured by the firm of Schaeffer,
Merkel and Company of Fleetwood, Berks County. He was informed on September 11, 1884, that the turbine was completed
and ready to be picked up. A dam was built across the road from
the new shop, and the turbine was installed that winter.

When the turbine was first started there was apparently some
difficulty in keeping the belt on the turbine drive pulley. A letter
from the company informed the Grubers that the main drive shaft
was eight inches too high, and they were advised to put a roller
on the top belt to keep it in the center of the turbine drive pulley.

The water turbine was small in comparison to the older type
water wheel. The Reliance Water Turbine was only twenty inches
in diameter. An outer casing of about thirty-six inches diameter
encased the turbine. Water, entering this outer casing from the
dam, submerged the turbine and was admitted to the turbine
through gates near the top. Figure 22 illustrates this construction.

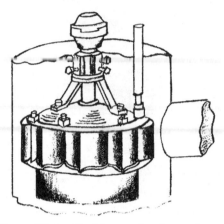

Figure 22. The Reliance Water Turbine and outer casing with the supply pipe on the right.

The control shaft extending upward from the outside diameter of the turbine opened and closed gates, which controlled the amount of water that entered the turbine and thereby controlled the speed of it.

The turbine was mounted vertically in a pit in the basement directly beneath the table saw in the wood shop. The control shaft extended upward through the floor next to the table saw where a wheel was mounted on it. This wheel was turned when there was a need to regulate the speed of the turbine. The drive shaft also extended up through the floor for no reason other than the Grubers did not wish to cut it off. Figure 23 shows the location of the turbine in the basement. The boards cover the pit that once contained the turbine. The wheel lying on the floor in the background was the jack wheel once used on the horse turnstile of the farm shop.

Steam Power

The Reliance Water Turbine served the power needs of the shop very well until 1896. By this time so much machinery had been added that the turbine was no longer able to supply enough power. To supplement the power source a steam engine was installed. An engine room was built on ground level in the corner where the wood shop wing joined the new building. A boiler

Figure 23. The location of the turbine in the basement. The large pulley on the left was driven by the turbine drive wheel.

room was added opposite the engine room at the outer corner of the wood shop. Figure 1, Appendix A, shows the location of these additions.

Gasoline Engine

The combined power of the water turbine and steam engine were sufficient until August 1906. At this time a combination of factors caused the Grubers to replace both with a gas engine.

The engine was an "OTTO" manufactured by the Otto Gas Engine Works in Philadelphia. Its serial number was 10337, and it delivered fifteen horse power at 260 RPM, with a governor to monitor the speed. The engine carried patents dated: July 31, 1888; August 21, 1888; September 26, 1888; December 4, 1888; August 5, 1889; August 27, 1889; September 2, 1889; October 24, 1893; and September 11, 1894. The engine was five feet long and one foot, nine inches wide. It had one cylinder that was six and three quarter inches in diameter. Its two flywheels were mounted on either end of a short crank shaft. These flywheels were four feet, six inches in diameter. The output drive pulley, which was mounted onto the same shaft, was seventeen and one quarter inches wide and thirty inches in diameter. The center of the drive wheel was twelve feet from the center of the main power drive line, which distributed the power of the engine throughout the works.

The engine would operate on either gasoline or kerosene; however, the Grubers always used gasoline. This engine delivered sufficient power for the shop until production was discontinued. The span of the power sources used in the Gruber Wagon Works was a direct reflection of advance of technology during the life span of the shop.

CHAPTER IV

MACHINERY OF THE WHEELWRIGHT

Until the beginning of the 19th century the hand methods of production by the craftsman were adequate to meet the demand for his product. At this time the efforts of hand production became insufficient. As the industrial revolution advanced, several factors of production improved and became available. Improved sources of power, a system of mass production, and improved machinery came into being. Usher intimated that the latter was most influential in the advance of industry.[18] This also appears to be the main factor in the development of the Gruber Wagon Works.

THE FARM SHOP

Early machines made by man were constructed mainly of wood, and this practice still persisted at the time F. H. Gruber established his farm shop. Consequently the few machines he had were made of wood.

Wood Turning Lathe

Because of the wheel hub the lathe became an important machine to the wheelwright at an early date. Mercer illustrates a wood turning lathe known as "The Great Wheel."[19] Sturt relates of a similar lathe used in his grandfather's shop in 1884.[20] Both of these machines were powered by a large wagon wheel which

---

[18] *Ibid.*, p. 328.

[19] Mercer, *op. cit.*, p. 222.

[20] Sturt, *op. cit.*, p. 57.

was turned by two men who operated the cranks attached to either side of the hub.

There was no information recorded on the method used to power the wood turning lathe in the early farm shop; however, Mr. F. H. Gruber recalled his father telling him that the lathe was at one time a "kick" operated lathe before it was powered by the horse turnstile. Later this same lathe was driven by the various power sources used in the new shop.

The lathe consisted of heavy oak legs and ways. The cast iron power head had a three step wooden pulley, which provided a means for varying the speed at which the work was rotated. The tailstock was mounted on an oak block, which was provided with a clamping screw operated from the bottom of the lathe. The tool rest, also of metal, was adjustable in height and proximity to the work.

The lathe was eight feet long and ten inches wide. Because there was a lack of space in the shop, the Grubers mounted a spoke cut-off saw on the lathe bed. Figure 24 reveals the construction of the lathe. On the wall behind the lathe are mounted

Figure 24. Wood turning lathe.

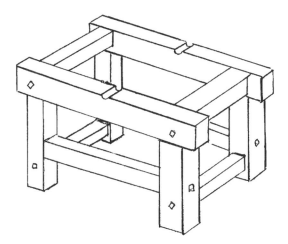

Figure 25. Base of table saw used in farm shop.

adjustable gages that were set to the dimensions of the hub being turned and used to check the diameter at the critical locations. This eliminated the use of the outside caliper and the need to stop the lathe to check the work. The cut out portion of the lathe bed permitted the downward movement of the drive belt at a previous location.

Table Saw

No record was made of the date that F. H. Gruber acquired it, but early in the operation of his farm shop he bought a table saw. It was made entirely of wood except for the shaft and saw blade. The base consisted of heavy oak legs and cross members that were bolted together. The drive shaft rested in two notches cut into the top cross members, and the drive pulley extended out to one side of the machine. Figure 25 indicates how the base was constructed. The top, also of wood, contained a slot through which the saw blade extended. It was hinged to the base at one end, and the depth of the saw cut was adjusted by raising or lowering the top with blocks of wood. This machine was moved to the new shop and served well until replaced by an improved saw.

Band Saw

On June 2, 1882, F. H. Gruber ordered a #3 band saw with extra blade from Frank & Company, Woodworking Machinery, 176 Terrace Street, Buffalo, New York. The bill, dated June 17, 1882, was itemized as follows:

> Frank & Co., Woodworking Machinery
> 176 Terrace St., Buffalo, N. Y.
>
> June 17, 1882.
>
> One No. Eureka Band Saw   135.00
> One 1 in. saw extra   4.70
>
>   recd payment   139.70

The Eureka Band Saw stood nine feet high. The main frame was a large cast iron "C" standing vertically. At the top and bottom, on center, were the two 40-inch drive pulleys. The drive belt pulley extended from the rear of the saw frame. The saw as it was received had absolutely no covers over the working parts; both the wheels and saw blade were completely exposed. The Grubers soon covered these exposed parts with wooden casings.

Tension was maintained on the saw blade by an adjustment of the top wheel. The adjusting handle was mounted beneath the curvature at the top of the "C" frame. This arrangement apparently was not satisfactory because the Grubers added a lever arrangement to increase the tension. It can be seen at the top of the saw in Figure 26.

The blades were sharpened by hand for some time; however, a hand operated saw filing machine was acquired shortly after moving into the new shop.

THE NEW SHOP

These three simple machines satisfied the needs of F. H. Gruber in his small farm shop; however, as soon as he moved into his new shop he started searching for more machinery. In the summer of 1885 he found four pieces of machinery he could use and purchased them. A bill indicated who he bought them from and the prices he paid:

Figure 26. Eureka Band Saw purchased June 1882.

The Reading Tack Co.
M. Harbester, Pres.
H. C. England, Treas.

Rdg., Pa.   Aug. 18, 1885

Shapeing Machine	$ 75.00
Jointer	75.00
Planner	100.00
Boreing Machine	25.00

Belts Puleys Hangers & shafting at half price
including counter shaft

The Reading Tack Company was located at 5th and Washington Streets, Reading, at the time of the sale.

Boring Machine

The boring machine was the oldest in design if not in actual age. It was constructed entirely of wood, except for the tool shaft and supports. It stood ten feet high. The vertical structure consisted of two three-by-six-inch oak beams reaching from floor to ceiling. The vertical tool shaft was mounted in the two cast iron hangers with sleeve bearings. The shaft was operated by a counterweighted lever which was fulcrumed by a pin driven through the three-by-six vertical members near the top. The boring shaft could be moved downward by pulling on the overhead linkage, and it was moved up by the heavy counterweight that was suspended on the rear of the operating bar. Figure 27 shows the details.

The table could also be moved up and down by a foot pedal. In this case the boring shaft was locked in position and the work was mounted on the table, which was moved upward until the boring tool had penetrated to the desired depth. The table could be tilted to a forty-five degree angle to the left or right.

This machine was used to bore holes in the draft pole and coupling pole, and it was also used to cut the round tenons on spokes with the tenoning tool shown in Figure 11. To do this the block was removed from the foot operated pedal. The tool was mounted so that it extended down through the hole in the table. The wheel was stood on edge and one spoke at a time was tenoned.

Jointer

The jointer, manufactured by L. Power and Company of Philadelphia, had a heavy cast iron base. The cast iron table was five feet eleven inches long and thirty-four inches wide. Both the front and rear parts of the table could be adjusted in height. Three adjustable cutting knives rotated on a horizontally rotating shaft. The drive pulley extended to the rear of the machine. The guide on the top of the table was adjustable so as to control

Figure 27. Wooden boring machine purchased from the Reading Tack Company in August 1885.

the width and angle of the cut. A safety guard covered the rotating cutter blades of this machine. The jointer can be seen in Figure 28. There was no provision for varying the speed of this machine such as a step pulley. It was driven by a belt from the main power line. To start it, the belt was shifted by lever from the idler pulley to the drive pulley.

Planer

There was no inscription on the planer to indicate who manufactured it. It was constructed of cast iron and had two separate

Figure 28. Jointer purchased from the Reading Tack Company in August 1885.

drive pulleys. One provided power to the table feed, and the other powered the cutting blade shaft. The feed rate could be varied by shifting the belt on the three step pulley provided, while the planer blades rotated at a fixed speed. The heavy cast iron table could be raised or lowered to accept lumber up to six inches in thickness. The ratchet-positioned handle on the left of the machine as seen in Figure 29 controlled the table height.

Shaper

At the same time that Mr. Gruber acquired the planer, boring machine, and jointer, he also purchased a table shaper. This machine was made entirely of wood as was the boring machine. It was thirty-five and one half inches high. The top, which was forty-seven and one half inches wide by forty-eight inches long, had a six-inch circular hole in the middle of it through which the tool shaft extended upward. The legs and cross members were heavy oak. The drive pulley was a three-inch diameter wooden cylinder about eighteen inches long. The tool was raised or lowered by an adjustment screw in the base, as shown in Figure 30.

Figure 29. Planer purchased from the Reading Tack Company in August 1885.

This machine was used to cut chamfers, shape small radius curves and other irregular-surfaced pieces of wood. Because it was made of wood, it vibrated considerably when heavy oak pieces were being worked, and the Grubers were continuously searching for a replacement for it. Another table shaper was acquired from the Gruber Organ Works of Stouchesburg, Berks County. It was also of wooden construction, and was not firm enough to handle the work that the Grubers wanted to use it for. It was therefore not used very long. Figure 31 reveals its construction.

In 1911 the Grubers finally found a shaper that they considered heavy enough to handle the work they were doing. It consisted of a cast iron base and heavy cast iron table. The tool shaft could be raised or lowered by a handle that was located beneath the right side of the table.

This machine was capable of doing the same type of work that the wooden types did. Through the use of templates it shaped curved pieces and chamfers. It was also used to shape the foot of the spoke. Figure 32 illustrates its construction.

Figure 30. Wooden table shaper purchased from the Reading Tack Company in August 1885.

Figure 31. Table shaper acquired from the Gruber Organ Works, Stouchesburg, Berks County.

Figure 32. Shaper acquired in 1911.

Mortising Machine

By 1894 the wagon works was in full production; however, the sons John and Jacob felt that there was a holdup in the production of hubs that had to be mortised by hand. F. H. Gruber insisted that this was a job that had to be done by hand in order that the fit be accurate. The sons, however, feeling that the use of a machine would speed up production and still do a satisfactory job, bought a mortising machine. The machine, serial number 5488, was manufactured by Goodell & Waters of 3101 Chestnut Street, Philadelphia. Its main structure was a cast iron pedestal that stood ten feet high. Figure 33 illustrates its appearance. A foot pedal was mounted at the bottom of the pedestal. The table on which the work was clamped was mounted in vertical ways that permitted it to move up and down. The table height was adjusted by means of a screw in the pedestal beneath the table. There were two tools mounted side by side in two separate spindles. Figure 34 shows the two tools.

The mortise was shaped in two steps. First the hole was drilled out with the boring tool. In the boring operation the drill

Figure 33. Mortising machine purchased in 1894.

did not move up or down. The table was moved up and down through the use of the foot pedal. The chisel squared the corners and sides of the mortise. As the chisel was moved up and down it was rotated ninety degrees to the right by a camming action built into the chisel spindle.

This machine was used to cut mortises into the parts of the body and carriage that required them. Another machine that had been dismantled and sold by Mr. F. P. Gruber had been used to cut the mortises into the hubs of the wheel.

Figure 34. The drill and chisel of the mortising machine.

Spoke Making Machine

In 1898 the Grubers decided to get a spoke making machine in order that they could make their own spokes for special jobs. They finally purchased a machine from Harry S. Bard Wagon Works, located at 824–26 Buttonwood Street, Reading. The machine was used for a time, but it proved to be unsatisfactory. It was mounted on a wooden base that was weak and caused considerable vibration when in operation.

In 1900 this machine was replaced by a machine that they purchased from a coach shop in Ashland, Schuylkill County. The machine had been manufactured by John Gleason of Philadelphia and was dated 1873.

It was all metal in construction and capable of making various objects such as pick handles, axe handles, and spokes through the use of patterns. The pattern was mounted in the upper chuck, and the stock from which the work was to be shaped was mounted in the lower chuck. Both chucks were rotated together, and as the follower traveled the length of the pattern the cutters

Figure 35. The spoke making machine purchased in 1900.

shaped the object from the stock. Figure 35 reveals the appearance of the machine.

The cutting head contained eight gouge-shaped blades mounted alternately on opposite sides of the rotating wheel. It revolved at high speeds as it traversed slowly left and right with the follower, which was held against the pattern by a strong spring tension. Both the pattern and the stock rotated slowly as the cutter traversed, thereby permitting the cutter to shape the stock as the pattern was. Figure 36 shows the cutter head and how the pattern is mounted in the machine. The lever in the left foreground was used to release the follower tension on the pattern and stock so that either could be replaced.

Spoke Tenoning Machine

In order to produce spokes at the same rate as other parts of the wagon were being produced the Grubers bought a tenoning machine in 1902. The machine, which had been made by the H. B. Smith Machine Company of Smithville, New Jersey, had a patent date of January 23, 1866. It had a cast iron stand with the

Figure 36. A close-up view of the cutter blades, follower and pattern on the spoke making machine.

power head mounted on the left side of the machine. On the stand were mounted two knife-edged tracks that permitted a tray-like bed to move from the back to the front of the machine. The tray had an adjustable guide to which the spoke was clamped by a lever. Figure 37 shows the details of the table.

There were two cutting heads mounted one above the other. They were adjustable to control the thickness of the tenon. A crank on the top of the power head controlled the distance between the cutting heads. Each cutting head had two blades mounted in it. The blades were sharpened on a bevel and ground at an angle; however, they were so mounted that their cutting edges were parallel when they were cutting the tenon. On the face of each wheel another set of blades was mounted. These blades were positioned so that they cut a sharp shoulder on the tenon. Figure 38 reveals the two cutter heads.

GRUBER-BUILT MACHINERY

At the time the Grubers were seeking machinery for their shop few machines were being offered on the market. Companies

Figure 37. Tenoning machine bought in 1902. The clamp can be seen on the left of the traversing table.

Figure 38. The power head of the tenon cutting machine. The crank on the top set the distance between the heads. The piece of tin over the cutters deflected the chips.

appeared to be reluctant to produce much machinery that apparently was needed. There appeared to be several justifiable reasons for this. One reason was the rapid rate of obsolescence of the machinery at that time. This was due to the rapid rate of innovations that were occurring. The other reason was the laxity and lack of sound patent laws. Strassman pointed out these facts in his discussion of the development of machinery in the 19th century:

> Nevertheless, the rate of obsolescence for a particular machine could not be predicted, and an enterprise built on a single novel machine was necessarily in danger.
>
> This danger was enhanced by the ease of circumventing patents and the ease of entry. The basic elements of metal working machines had almost all been developed before the end of the eighteenth century. When a new metal working machine became commercially feasible, it could often be constructed with several different combinations of these basic elements. The number of patents a man held were therefore more a mark of prestige than a mark of prosperity. No doubt patents delayed competition somewhat, but a British observer found that "a shameful copying of everything which reaches a successful sale" was the rule in the United States.[21]

It was apparent that many people did not consider the patent laws effective. "Some innovators, among them Eli Whitney and William Bement, did not even find patents worth the effort and expense of their acquisition."[22] Following the accepted practice of the times, the Grubers built many of the machines that their work required. They used those elements of machines that they were familiar with and recombined them into machinery that filled their needs, often working into the night to do so.

---

[21] Strassman, *op. cit.*, p. 148.

[22] *Ibid.*, p. 149.

Dado Cutting Machine

Many parts of the Gruber Farm Wagon were assembled through the use of the dado joint. To make this joint by hand methods required the skills and much time of an expert craftsman. In order to circumvent the need for the expert craftsman and to save time while still achieving the quality of fit they wanted, the Grubers devised the dado cutting machine.

This machine, like many others up to this time, was made of wood. To give it the strength and rigidity needed, it was made of heavy oak. The base provided a firm platform on which was mounted two parallel knife-edged tracks for the top to slide on. The top was adjusted by the means of a wheel at the lower edge of the base, which moved the top up or down so that the cut made by the tool could vary from one to two inches in depth. The top could also be adjusted to the left or right to control the width of the cut. The large guide, to which work was attached by the means of a clamp, was also adjustable to various angles as needed. Figure 39 shows the construction of the machine.

Figure 39. Dado cutting machine made by the Grubers.

Figure 40. The cutting head of the dado cutting machine shown removed from the shaft.

The drive shaft was mounted in metal bearing plates on the top of the base so that it was perpendicular to the direction in which the top moved when it was pushed back and forth on the tracks. The cutting head consisted of two bars of tool steel mounted on the shaft so that the four ends were ninety degrees apart. Each of the four ends of the cutting head contained two cutting bits. One chipped the wood and shaped the bottom of the cut while the other tool bit cross-cut the wood and formed the end of the dado. The relative position of the two bars could be changed to regulate the width of the cut made from one inch to two inches wide. Figure 40 shows the cutting head removed from the machine.

This machine was used to prepare the various parts of the carriage for assembly. Dados were cut into hounds, axle beds, bolsters, and bolster beds. The locations of these various dados were marked through the use of patterns.

Figure 41. Spoke tenoning machine built by the Grubers.

Spoke Tenoning Machine

The round tenon at the head of the spoke could be shaped by various means. The older wheelwrights at one time cut them by hand with the saw and draw knife.[23] Mr. Gruber used this method when he first started making wagons. Later, as tools developed, he used the tenoning tool. This method, although faster and somewhat more accurate, was still not satisfactory for the demands of the Grubers. In response to these demands they built a spoke tenoning machine, as shown in Figure 41.

This machine was also built mainly of wood. The base was made of heavy oak members. The top of the base had no cross member on the left end. The wheel was mounted on two adjustable sets of blocks in which the axle was clamped. Mounted in this fashion, the wheel could be centered on the tool and rotated so as to present each spoke to the tool.

Figure 42 shows a spoke about to be clamped into the cutting position. The drive and operating mechanism was mounted

---

[23]Jenkins, *op. cit.*, p. 67.

Figure 42. Wheel mounted in spoke tenoning machine in readiness to have tenons cut.

on the right end of the machine. The shaft on which the tool was mounted was directly in line with the axis of the wheel. This placed the spoke in line with the cutting tool, and as the shaft rotated the operating lever was pulled to the left drawing the tool onto the spoke end. The tool head consisted of a solid disc on which was mounted the four cutting bits. Two bits shaved the diameter of the tenon and the other two cut the shoulder of the tenon. Different heads were used for the various sizes of tenons, although the depth was regulated by adjusting the stroke of the operating lever.

The machine was adjustable to accommodate various sizes of wheels. To do this the top of the table was moved to the left or right. The index marks can be seen in the edge of the base in Figure 41. They were graduated according to the diameter of the wheel, and were set by moving the top index to the diameter of the wheel to be worked on.

There was another tool used to trim the spokes to the proper length. It was a simple circular saw which could be mounted in the same shaft as the tenoning tool. Whereas the spoke was held in the double "V" jaw chuck when the tenon was being cut, it was rested on the top of a swinging block to place it in the proper position to cut it off.

This machine was also used to true the diameter of the wheel after the felloes were assembled. The tool used for this operation was a disc with a shaving bit mounted on its face near the perimeter. It was mounted on the rotating shaft in the same manner as the tenoning tool; however, the wheel was mounted in a higher set of bearing blocks so that the tool would be in line with the axis of the wheel as it rotated. The set-up may be seen in Figure 43. After the wheel was mounted in place it was rotated, and the rotating tool shaved the felloe to the true diameter.

This machine not only enabled the Grubers to reduce the amount of time that it took to make a wheel by the slower hand method, but also enabled them to make a better wheel. They were not only more uniform, but also they were better fitted and therefore, less likely to work loose later.

Figure 43. Wheel mounted in position to cut the true diameter on the spoke tenoning machine.

Curved Surface Shaper

Many of the parts of the wagon carriage, such as the axle beds, bolsters, bolster beds, hounds, and sliding bar, were shaped with curves. Some were due to function, some due to a need to reduce weight, for oak was a very heavy wood, and there was much of it in a wagon. These curved surfaces were at one time finished with the draw knife. F. H. Gruber in his early shop used the band saw and sander to shape them. In order to expedite the process the Grubers bought a shaper in July of 1892. The following bill indicates the price and source:

> July 14, 1892
> to F. H. Gruber
> from Geo. W. Francis
>   Practical Machinist
>   126 Carpenter Street
>   Reading, Pa.
>
> To one Shaping Machine $30.00
> wooden base, two blades, discs and clamps.

This machine was constructed of wood and was, according to Mr. F. P. Gruber, too light for the work it was needed for. Some time later John Henry Gruber, the brother of F. H. Gruber, and the sons rebuilt the machine. They made patterns and had castings made for the base and foot pedals. In this they assembled the original blades, discs, and clamps. Later the frame was covered on the operator's side in order to prevent the chips from flying into the face of the workman who had to hold the clamping lever down with his foot. The machine is illustrated in Figure 44.

The cutter head consisted of two blades mounted in a horizontal shaft. On either side of the cutter head was a guide disc of the same diameter as the cutter head. These were backed by two guide plates. The movable clamping head was mounted over the cutter head.

This machine was used to shape the rough-cut piece to the shape of a pattern. The pattern was placed on the rough-cut shape and held there by the gugs in the pattern. The pattern was placed

Figure 44. Curved surface shaper after it was rebuilt by the Grubers.

against the guide plate according to the grain of the wood. The clamp was placed against the wood to hold it there. The piece was slowly pushed down so that the pattern rested on the guide disc. It was then drawn through the shaper as it was held down on the guide disc. If the grain of the wood changed, the piece had to be turned around so that the pattern was held against the guide plate on the other side of the machine. Generally only one

Figure 45. Wooden belt sanding machine used to finish rough-cut surfaces of undercarriage parts.

pass through this shaper was needed to complete the job to the satisfaction of the Gruber craftsmen.

Sanding Machine

In the early shop, when work was done on a piecemeal basis, there was little need to sand the work. The drawknife, spoke shave and plane finished the surface to the satisfaction of the craftsman. Later when the band saw came into use F. H. Gruber used the sanding machine to finish the surfaces of the product.

The first sanding machine was made of wood. The base was seven feet long and eighteen inches wide. The belt traveled on two wooden drums, each of which was ten inches wide and eight inches in diameter. Tension on the belt was maintained by a large screw threaded through a block attached to the end of the base and connected to the movable rack in which one of the drums was set. Figure 45 shows how this was done.

This wooden sander soon became inadequate, and the Grubers designed and built another sanding machine. This new

Figure 46. Belt sanding machine built by the Grubers.

machine was made from cast iron parts. It consisted of a pedestal with a six-foot adjustable bar extending from the top. The length of this bar was adjustable to maintain tension of the sanding belt. As on the wooden sander there was no back-up beneath the sanding belt. The work was sanded over the floating belt. This sander had one advantage for the operator that the older one did not. It was positioned next to a door, which could be opened when the sander was in use and the dust dissipated outside the building. Figure 46 shows the position of the sander.

Table Saw

The wooden table saw was the next machine to be replaced by Gruber-built machinery. The older saw became inadequate for the work load that was being placed on it, and the Grubers sought a newer one to replace it. Being unable to find one, they decided to build one.

This new saw was made of cast iron. It was cast in patterns that the Grubers made. It was cast in a foundry in Reading and returned to the shop to be machined and assembled. The base had solid cast iron legs and frame. The top, likewise, was of cast iron. It was also adjustable so that the depth of the saw cut could be controlled. Figure 47 illustrates the construction of the saw.

This saw did not have the standard rip and crosscut guide, as it was used mainly with special jigs made by the Grubers to cut standard parts. A rip fence could be set on the table by using a slot across the top, and a crosscut guide could be used in the slot running the length.

This saw was used primarily to cut the pieces of lumber used in the manufacture of the bodies of the wagons. Each part was cut in a special jig so that each would be identical. These jigs were hung on the walls around the saw. A twenty-foot-long jig used to cut the boards for the body was stored by hanging from hooks in the rafters over the saw.

No provision was made for the saw to be tilted as there was no need for compound cuts. All cuts made by this saw were

Figure 47. Table saw made by the Grubers. The crank on the left set the depth of the saw. The lever in the left foreground shifted the belt to engage the drive belt.

either cross or rip cuts, as these were all that were needed to make the Gruber Farm Wagon.

Hub Boring Machine

Difficulty in one of the processes in making the wheel caused the Grubers to seek a machine to improve their technique. The hand methods of boring the hole in the hub were satisfactory when the axles were wooden and there was plenty of time to hand-fit the hub. However, with the coming of the iron axle, more precision was needed to make a wheel that would not wobble after use under load. For this reason the Grubers sought some improved technique by which they could bore the hole in the hub.

There were machines on the market designed for this function; however, they were hard to find, and the Grubers felt that they could make one that would work as well if not better than one they could buy. Again they made the patterns and had the parts for the machine cast in iron. The result was a machine that was capable of boring the hole in the hub for the bearing, bushing, and the nut in two simple operations. Figure 48 reveals the result.

This machine consisted of a cast iron base and stand. The operating shafts were mounted in bearings on either end of the machine in such a manner that their rotating axes were in line with each other. Each shaft had its own drive belt and operated individually. The boring tools were mounted on the ends of these shafts to bore the required size holes. The hole was bored by moving the operating lever either to the left or right depending on the side it was on. Figure 49 shows a wheel mounted in position.

In the center of the machine was a large circular cast iron frame. In the center of the frame a double "V" jawed chuck was mounted. The wheel was placed with its front side against this circular frame and the hub was clamped in the chuck, which held it firmly in place during the boring operation.

The Grubers considered this operation a critical one: so much so that they designed and made a special tool for each hole to be bored. These tools were stored in a cabinet mounted on the wall behind the machine. Figure 50 shows the cabinet with the tools and information posted in it.

Figure 48. Hub boring machine made by the Grubers. The operating levers may be seen on the extreme left and right of the machine.

Figure 49. A wheel mounted in the hub boring machine. The crank on the left operated the clamping chuck.

Figure 50. Cabinet containing the hub boring tools and the list of combinations to be used for various sizes of wheels.

Spoke Trimming Saw

The spokes that the Grubers drove into the hub of the wheel were not allowed to extend into the center hole. To prevent this, the foot of the spoke had to be trimmed at an angle. To perform this simple though precise operation, the Grubers devised a simple machine they called a spoke saw. See Figure 51. The base of this saw was an oak plank eight inches wide, twenty-four inches long and two inches thick. The rests on which the table was mounted were perpendicular to the length of the base. One was mounted at either end. On the rest near the saw blade a

Figure 51. Spoke trimming saw mounted on the ways of the wood turning lathe.

knife-edged track was imbedded to guide the tray with accuracy across the saw. The blade was mounted on a short shaft on the left end of the base. Figure 52 shows the position. On the front edge of the traversing tray a guide fence was mounted. Next to it, on the left, a small guide was positioned, onto which the shoulder of the spoke was set. This guide was set to control the length of the tenon. It was set according to the scale inscribed on the edge of the guide fence. The hind side of the spoke was set against the fence, and the cut was made.

THE BLACKSMITH SHOP

When the Gruber Wagon Works began production the blacksmith shop contained only the bare necessities — two forges, anvils, and an assortment of hammers, tongs, and small tools that enabled the blacksmith to shoe a horse, iron a wagon or tire a wheel. These simple tools, however, were not sufficient to enable the blacksmith to keep pace with the increased production that soon set the tempo of work in the rest of the shop. Machines were needed, and over a period of time they were acquired.

Figure 52. The saw blade and tray of the spoke trimming saw. The edge of the guide is inscribed with settings.

Power Punch and Shear

The largest, although not the most unusual, machine in the blacksmith shop was the power punch and shear, manufactured by the Royersford Foundry and Machine Company of Royersford, Pennsylvania. It was dated 1894 and identified as serial number 2, with a shop number 1438. It was second hand when the Grubers acquired it in 1905; however, it was in good condition and served the Grubers well.

Although the machine was capable of both punching holes and shearing metal, the Grubers used it mainly for the latter. Metal up to seven-sixteenths of an inch could be cut cold in the shears. This expedited the work of the blacksmith considerably when he had to cut stock in preparation for tiring wheels and ironing wagons.

The power punch and shear may be seen in the left background of Figure 53. Its massive cast iron structure stood ten feet high. At the top were two large cast iron flywheels, which added their impetus to the drive of the machine. The tools of the

Figure 53. View of the interior of the blacksmith shop looking toward the north.

machine were located low on the left and right side of the stand and were relatively small for the size of the machine.

Power Press

Putting the bearing in the hub, or "boxing the wheel" as the older wheelwrights called the process, was not as easy as it had been. Older wheelwrights used hand tools for this job. Some used the reamer to enlarge the hole. Sturt speaks of using chisels and gouges to shape the hole to receive the bearing.[24] With the advent of the iron axle and bearing, the process of boxing the wheel required more precision than previously applied. The bearing could no longer be wedged into place; it had to fit tightly in the precisely drilled hole prepared for it.

To do this job the Grubers purchased a power press from the Keystone Wagon Works, located on Buttonwood Street near Third Street in Reading. The press had been manufactured by the Defiance Machine Works of Defiance, Ohio.

---

[24]Sturt, *op. cit.*, p. 130.

Figure 54. Power press used to press bearings into the hubs of wheels.

The press was a mechanical type press that operated on the screw principle. The eight foot wide base angled up toward the center where the anvil was located. At either end two eight-inch-diameter columns rose to a height of ten feet. Across the top of these two columns was suspended the drive head and arbor of the press. Pressure was exerted through a drive pulley on the top of the right column.

To press a bearing into a hub, the wheel was placed hind side up on the anvil. The bearing was placed in the hole prepared to receive it, and the press was lowered until the bearing was fully in place. Figure 54 shows the structure of the press.

Hydraulic Tire Setter

The most interesting and advanced machine in the shop was the cold tire setter. It was interesting because of its design and function, and it was undoubtedly advanced because the principle of hydraulics was applied in its performance.

This machine was the most expensive ever acquired by the Grubers as indicated by the following bill:

Rochester, N. Y.   Dec. 2, 1908

The West Tire Setter Co.
mfr. of West Hydraulic Tire Setter and Hub Bander
509 Ellwanger and Barry Building

Second hand #3 Hydraulic Tire Setter #328
fitted with small control valve
                              Net $1000.00

This machine consisted of two parts: one was the hydraulic press, and the other was the pump that supplied the hydraulic pressure to operate it. The press was arranged in a circle that was five feet six inches in diameter. Around the perimeter were eighteen hydraulic pistons arranged in such a manner that their pistons were extended radially toward the center of the machine. Each piston was connected to the hydraulic pump by a pipe so that each received equal pressure. To adjust the setter in order to accommodate the various sizes of wheels the feet of the pistons were changed. These came in sets of different lengths.

The hydraulic pump that supplied the pressure stood to the rear of the setter. It was a two-cylinder arrangement that was driven by a large drive pulley. The valve that controlled the flow of oil was mounted between the two parts of the machine. Drip trays, designed to catch and return the oil that dripped from the pistons, were placed beneath each piston. Figure 55 illustrates the construction of the machine.

The unique aspect of this machine was that it pressed the tire on the wheel without having to heat it. The setter was prepared for use by placing the proper sized blocks in place with the pistons fully withdrawn. The wheel was placed in position

Figure 55. Hydraulic setter used to set tires on wheels. The pump is in the background.

hind side down on one-eighth-inch shims. These shims were used to position the tire so that it was centered on the felloe, in which case the tire extended one eighth inch on either side of the felloe. The tire, made oversized, was slipped over the wheel, and the pressure was applied to the pistons, which closed and compressed the tire on the wheel. This process saved many man-hours of labor in the blacksmith shop, and the Grubers advertised the process widely.

Form Roller

The forming of the bar of iron into a perfect ring to make the tire was at one time a difficult chore. Jenkins mentions that this was done in the older shops by bending the iron around a post.[25] Both he and Sturt recall that later shops used a tire bender to do the job.[26] The description given by Sturt is very similar to an older machine in the Gruber shop. Figure 56 shows this machine. It was operated by hand cranks, one on either side. The degree of

[25]Jenkins, *op. cit.,* p. 75.

[26]Sturt, *op. cit.,* p. 118.

Figure 56. Hand operated tire bender.

bend made by the machine was adjusted by the position of the side rollers. The rollers were placed close to the drive roller to make a small ring and out for a large ring.

A power form roller was added to the machinery of the blacksmith shop sometime after 1905. This machine was power operated and much stronger in construction than the older machine. Figure 57 illustrates the structure.

Tire Shrinker

Another machine in the blacksmith shop that was unique in its application, although not in its principle of operation, was the tire shrinker shown in Figure 58.

This machine was hand operated by the use of a large wheel mounted on the right side of the base. There were two self-locking, eccentric cam type clamps in which the wheel was held. One clamp was on the fixed part of the machine while the other was on a movable part. As the operating wheel was rotated the movable part of the table moved toward or away from the other. As they moved closer together the clamps tightened, and further closing caused the metal to upset and thereby "shrink" the

Figure 57. Power operated tire bender.

tire. This machine was used mostly on old tires that were re-moved from wheels that had been in use and became loose for some reason or other. This machine was manufactured by D. H. Potts of Lancaster.

Thread Cutting Machine

Because the wagon was assembled through the use of many bolts, the blacksmith spent much time cutting threads: sometimes on bolts he made himself, other times on bolts that were not threaded quite as much as needed.

There were several sets of stocks and taps about the shop; however, the Grubers felt the job could be done quicker and bet-ter with a machine. They had two machines, one hand operated and the other power operated.

The hand operated machine was a simple device. It consisted of a base and two uprights. In the right upright a wheel was mounted on a horizontal shaft. On the other end of the shaft was a clamp in which the stock to be threaded was held. The die was clamped in the left upright, and the stock was run into the die, thereby cutting a thread on it. This machine, made by the

Figure 58. Tire shrinking machine.

Neverslip Mfg. Company of Brunswick, New Jersey, is shown in Figure 59.

The power operated thread cutter, manufactured by Wells Bro's. and Co., of Greenfield, Mass., worked slightly differently than the hand operated machine. It looked much like a lathe, with a power head on the left end of the ways and a carriage that was moved left and right by a crank on the rear of the machine. The die was clamped in the rotating chuck, and the bolt stock was clamped in the vise on the carriage. The thread was cut by feeding the stock into the die as it rotated. This machine is shown in Figure 60.

Figure 59. Hand operated thread cutter.

Drill Press

There were two floor-type drill presses in use in the blacksmith shop. They were similar in design to contemporary machines of this type. A small bench-type drill press was unique in the simplicity of its design and function. It was manufactured by Howley & Hermanec of Williamsport, Pa.

An open, rectangular-shaped frame having two bearings on the upper and lower corners in front held a keyed shaft, which was the spindle. A pulley keyed to the shaft was held between the two bearings. An operating lever, fulcrumed in the middle on a projection from the top of the frame, was attached to the top of the spindle in the front and counter-balanced in the rear. A horizontal drive pulley was attached to the upper rear corner of the frame. This machine was mounted on a heavy post, and a plank mounted just below it served as a table. The drill had no chuck as such. A collar was pinned to the bottom of the shaft and the drill bits were fastened in this collar with a square-headed bolt. The press is shown in Figure 61.

Figure 60. Power operated thread cutter.

Metal Turning Lathe

The manufacture of the Gruber Farm Wagon did not require the use of a metal turning lathe. However, making and repairing the machinery, which the Grubers did, had that requirement. This lathe was placed in the engine room, although there was no apparent reason for doing so.

The features of construction were similar to contemporary lathes, except the parts are much heavier. The headstock, mounted on the right, contained the gearing. On the right end of the ways was the movable tailstock. The tool holder was capable of moving left or right and cross feeding. There was no provision for angle cuts; however, the tool mount did have an automatic feed to the left and right. There was no way of measuring the depth of the cut. The chuck was a three jaw universal type. The lathe is shown in Figures 62 and 63.

If this consistent application of ingenious innovations and constant search for better equipment could be construed as technological advancement, then the Gruber Wagon Works was an excellent example of such advancement.

Figure 61. Simple bench drill press.

Figure 62. Metal turning lathe.

Figure 63. Chuck, carriage, and tailstock of lathe.

CHAPTER V

PROCESSES

By the middle of the 19th century most of the basic elements of the industrial revolution had been introduced to American industry. Mass production, interchangeable parts, improved machinery, and better power sources were all being applied to the factory system of manufacturing to some extent.

Early Farm Shop

When he first began to repair and make wagons in a small shop on his farm, Mr. Gruber was not aware of, nor did he need, these new developments. He used the processes that were taught to him by his ancestors. Each wagon was manufactured individually, each being made when there was a need. As the demand for his services and products increased, Mr. Gruber began to seek new methods and processes.

As the work load became too great for him and his young sons to handle in his farm shop, F. H. Gruber employed the services of blacksmith Henry L. Fiant. However, because there was not enough work in the farm shop to support the full time services of the blacksmith, he worked in his shop in Wernersville. The wood for the wagons was prepared in the farm shop and then taken to the blacksmith shop where the wagon was assembled and "ironed."

The process of ironing a wagon included its assembly with the iron bolts, braces and pins; installing the iron axles in their beds; and placing the bearings, bands and tires on the wheels. This arrangement between the wheelwright and the blacksmith was difficult and time consuming. There were no telephones, so all communication was done through letters such as the following which was postmarked Wernersville, Feb. 26, 1883:

Wernersville Feb. 26, <u>1882</u>

Mr. Gruber your wagon will
be done tomorrow <u>after noon</u>
I would like if you would
<u>Fetch</u> it at <u>onsed</u> and bring
that other wood along for the
other wagon and if you have
some of that <u>light-wood</u> work
ready bring it along.

Yours Respectfully
Henry L. Fiant

This system not only required much correspondence, but it was also time consuming in other ways. The wagons and materials had to be transported to and from the blacksmith shop. This was done by one of the sons, and it was a half day trip by wagon. Technical problems also caused delays. Sometimes Mr. Fiant was not quite sure what Mr. Gruber wanted done, and other times the opposite was the case. Either way this created the need for more communication such as the following:

Wernersville Aug. 29, 1882

Mr. Gruber If you can, come over on <u>thursday</u> and
bring <u>them</u> other <u>set</u> of <u>Wheels</u> along the <u>reson</u> I
want you to come is about the tongue pole I do
not like to cut it down to put the chains on
for <u>feer</u> I might spoil it and that I would not
like to do and if you come bring your <u>Blain</u> along
and drawing knife that is all you need We <u>doned</u>
get this wagon <u>quict</u> done this <u>weeke</u> but
the <u>reson</u> I want them other <u>Wheels</u> is because
Rollman is going to <u>lieve</u> me till <u>weeke</u>
after next Then I am going to put the tires on
and <u>welled</u> the axles before he <u>lieves</u> so
that we can go ahead anyway I know another

Blacksmith but I due not no Wether I can get
him right away or not but therefore I will try
to arrange it so we can go ahead but if you think that
you can Finish that pole as easy after the pole is put
in you can do it Friday or saderday that makes no
difference to me but I thought it would be better
before it is in the wagon

> Yours
> H. L. Fiant

This arrangement with Mr. Fiant had other disadvantages,
one of which was the size of the blacksmith shop. Because it was
a small shop, the blacksmith could not keep the wagons inside
the shop when they were completed. This caused him some con-
cern, as the following letter indicates:

Wernersville, Nov. 22, 1882

Mr. Gruber Your wagon is done and them other
wheels are done too; and will you please fetch it as
soon as you can as we have a bad place to have it
stand on account of horse shoeing as it is not very
nice to shoe outside by this kind of weather.

> Respectifully Yours
> Henry L. Fiant
> per D. Crumbine

The New Shop

The new shop, known as the Gruber Wagon Works, was
arranged in the manner of a small factory. It was a three story
frame structure which was divided into separate work areas.
Appendix A gives details of the structure. The power sources
were placed in the sub-ground floor area where the main drive
line, which was suspended from the first floor joists, distributed
its energy throughout the building. The first floor was divided
into three areas. One was the blacksmith shop, where the iron
work was done; the second was the wood shop, where wood
was made into the various parts used in making the wagon; and
the third was the bench shop, where hand work was done, and

part of the wagon was assembled. The second floor contained two areas. One was the paint shop where the wagons were painted, and the other was a storage area where wagons were stored until they were sold. The third floor offered more storage space, in which the precut parts of the wagon were stored until needed.

These were the physical facilities within which the Grubers applied the various elements of industry as the need arose. They changed and increased the source of power they used. They increased and improved the machinery they used. Last but not least they improved the processes by which they manufactured their main product - the Gruber Farm Wagon.

THE WHEEL

Of all the parts of the wagon, the wheel was undoubtedly the most difficult to make. The cutting, shaping, and assembling of the hub, spokes, and felloes taxed the skills of the wheelwright to the utmost. When F. H. Gruber moved into his new shop he continued to make his wheels entirely by hand; however, within a few years his sons were applying several machine operations to the process.

Hub

In making a hub the Grubers started with a twelve-to-fourteen-inch-diameter log, cut into ten-to-fifteen-inch lengths. These were drilled through the center with a two-inch drill. Although Sturt regarded boring the hole in the hub at this time with disfavor, it seemed to be the common practice to do so at this time.[27] Most wheelwrights felt that it accelerated and aided the drying of the wood. The wood was then placed in storage for four to five years in order to allow it to dry thoroughly.

After the wood was properly seasoned and ready to be used, it was mounted in the lathe through the use of a long tapered spindle, which was pressed into the hole previously drilled through the hub. The diameter of the hub was now turned to the

---

[27]Sturt, *op. cit.,* p. 48.

Table 1. Data on Marking Gages According to Hub Sizes for Various Axles.

Axle Size	Hub Size	A	1	B	2	C
2 1/4"	10 x 13"	1 7/8	3 1/2	2 5/8	3 1/2	1 1/2
2"	9 x 12"	1 7/8	3 1/8	2 3/8	3 1/8	1 1/2
1 7/8"	8 1/2 x 11"	1 5/8	3	2 1/8	3	1 1/4
1 3/4"	8 x 11"	1 1/4	3	1 7/8	2 5/8	1 1/4
1 5/8"	7 1/2 x 9"	1 1/8	2 3/4	1 5/8	2 5/16	1 1/8
1 1/2"	6 1/2 x 8 1/2"	1 1/8	2 5/8	1 1/2	2 1/8	1 1/8
on back of above			3 9/16	1 5/16	3 3/16	
1 3/8"	6 x 8"	1	2 9/16	1 3/8	2 1/16	1
1 1/4"	5 1/2 x 7 1/2"	1	2 3/8	1 1/4	1 15/16	15/16
	12 x 16"	1 3/4	4 1/2	3 1/8	4 1/4	2 3/8

desired size and marked with a gage to identify the location of the various cuts to be made next. These gages were made from pieces of scrap wood about eighteen inches long, two inches wide and one quarter inch thick. Small brads were driven into the edge of the wood at the points that were to be marked. The brads were then sharpened to a point so that they would scratch a clean line on the diameter of the hub as it was rotated in the lathe and the gage was held against it. Table 1 shows the data given by gages.

Pins were placed to mark the distance for spaces marked a, 1, b, 2, and c. The marking gage is shown in Figure 64. The gages were hung on a string through a hole in the gage.

The distance marked by "a" was for the front band. The dimension marked by "1" gave the distance between the front band and the mortise, while "b" gave the width of the mortise. Dimension "2" was the distance from the mortise to the hind band and dimension "c" was the width of the hind band.

After the hub was marked it was turned to the desired diameter. These diameters varied with the size of the hub. The data for these diameters were posted on a chart located on the door of a cabinet hung on the wall behind the spoke tenoning machine. The data are shown in Table 2.

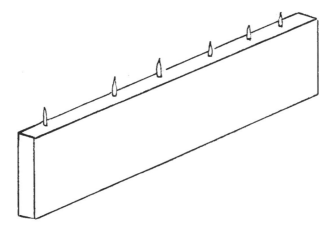

Figure 64. Marking gages used to locate and mark the dimensions of the hub.

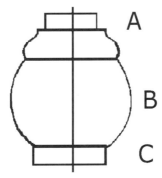

Table 2. Diameter in Inches of the Hub at the Points Specified.

AXLE SIZE	A	B	C
2 1/4	7 1/2	9 7/8	8 3/4
2 1/8	7 1/8	9 3/8	7 7/8
2	6 3/4	8 7/8	7 3/8
1 7/8	6 1/4	8 1/4	7
1 3/4	5 3/4	7 3/4	6 3/8
1 5/8	5 1/4	7 1/8	6
1 1/2	4 3/4	6 5/8	5 1/2
1 3/8	3 7/8	5 7/8	4 5/8
1 1/4	3 5/8	5 1/4	4 1/4

Rough-turned hubs were also bought from commercial sources. These hubs were already drilled and turned to an approximate diameter. They were mounted in the lathe the same as the homemade hubs and marked the same way. Figure 65 shows two types of hubs that were bought. One was plain while the other was rough-turned. Shown in Figure 66 is a hub mounted in the lathe with the marking gage in place as for use.

After the hubs were turned to the specified size and shape they were mounted in a jig on a hub mortising machine. An indexing head was set for the diameter of the hub at the point where the mortises were to be cut. In the older shop these mortises were layed out and cut by hand; however, as the work increased the sons convinced F. H. Gruber that a mortising machine would do the job faster and just as well.

The edges of the mortises were bevelled to a forty-five degree angle to insure a better fit of the spoke. Then the hub was taken to the blacksmith shop where the mortise bands were driven into place on either side of the mortises. These bands, generally one eighth inch thick and one inch wide, were positioned to prevent the hub from splitting when the spokes were driven into place.

Spokes

Of all the parts of the wagon, the spokes of the wheel were undoubtedly the most demanding of care and exactness. The Grubers used various means in making spokes. Sometimes they made them entirely by hand; sometimes they used the machine process; and at times they purchased ready-made spokes.

Before they had the spoke making machine, and when they could not buy the spokes that they needed, the Grubers made the spokes by hand. This process started with the careful selection of oak or hickory logs that were free of defects and were straight of grain. They were cut into lengths, split into suitable sizes with an axe or frow, and then put away to dry. At times there were as many as two wagon-loads of spokes processed in this manner at one time. Figure 67 (top) shows a piece of wood prepared for drying.

Figure 65. Two types of commercial hubs bought by the Grubers.

Figure 66. The hub marking gage in use.

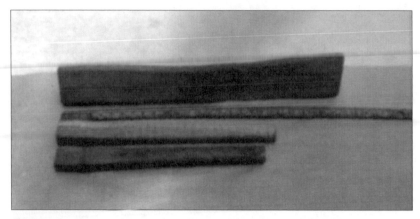

Figure 67. Three sources of spokes. At the top is a split of hickory from which a spoke will be make by hand. In the center is a rough spoke as it is removed from the spoke making machine, and at the bottom is a commercial spoke.

When spokes were needed, sufficient stock was taken to the bench shop, where the process of making the spoke began. The piece of wood was first planed to the rough size and shape of the spoke. Care was taken so that the grain of the wood ran with the widest part of the spoke. The Grubers felt that the spoke was weak when the grain ran across the width of the spoke. Figure 68 shows the proper way the grain should run.

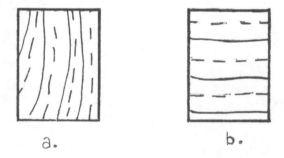

a.   b.

Figure 68. End grain of spokes: (a) shows the grain of a correctly shaped spoke, while (b) shows that of an incorrectly shaped spoke.

The spoke was next put in a clamp on Jacob's bench as shown in Figure 69. Here with the drawknife and spoke shave the spoke was shaped. The foot or rectangular end of the spoke that was inserted into the hub was later shaped with the drawknife.

Another method of making spokes was a machine process. When using the spoke making machine the Grubers found that they could save time and labor by sawing rather than splitting the wood in preparation for shaping the spoke. According to Jenkins, the English wheelwright frowned on this process:

> No other part of the wheel bears greater pressure than the spokes and for that reason only well seasoned cleft heart of oak is suitable for spoke making. Although some modern spokes have been made of sawn oak they are far from satisfactory for the grain must be unbroken and the cleavage must follow the grain.[28]

However, the Grubers thought this method satisfactory as long as they exercised caution in sawing the wood. The sawed

---

[28]Jenkins, *op. cit.*, p. 66.

Figure 69. Spoke clamped on wheelwright's bench.

piece was placed in the spoke making machine with a pattern and cut to shape.

This machine made what was known as a club spoke or a spoke without a tenon, as shown in Figure 67(center).

Factory-made spokes were also supplied in this shape, although they could be bought with a semi-finished foot, as shown in Figure 67(bottom). The Grubers purchased many spokes due to the lack of facilities and time in the old shop. At the time they bought mostly from local sources such as H. S. Bard at 822, 824, and 826 Buttonwood Street, Reading. They usually bought the best grade available, which was generally oak, unless the customer specified otherwise. An indication of their insistence upon quality material is given in a letter from a supplier:

> Bard Hardware Co.
> 800 Penn St., Reading, Pa.
>
> Gruber Wagon Works
> Obolds, Penna.
>
> Gentlemen:
>
> As you requested we are prepared to quote you the following prices on spokes namely:
> 1 5/8 A grade best quality 6.50 per set   Plain
> 1 3/4 A grade best quality 7.35 per set   Hickory
> 2 A grade best quality 8.85 per set   Spokes
>
> 1 3/4 sec growth best quality 5.25
> 2 sec growth best quality 7.00
> 2 1/4 sec growth best quality 8.00   Plain Oak Spokes
> 2 1/2 sec growth best quality 9.00
> 2 3/4 sec growth best quality 11.00

The next step in making spokes, except the handmade variety, was the cutting of the tenon on the foot or the rectangular end of the spoke. This was done on the tenon cutting machine. The spoke was clamped in the tray, the foot was passed between the cutting heads, and the tenon was cut to the proper thickness.

The spoke was then taken to the tenon cut-off saw on the lathe bed, and the tenon was cut to the proper length and angle.

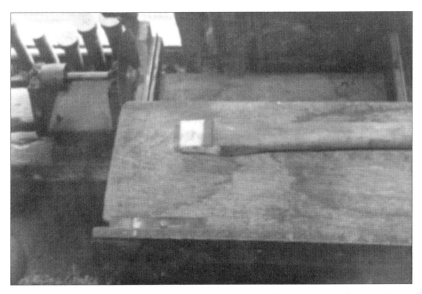

Figure 70. Foot of a spoke being cut at an angle to permit the spoke to assume the dish angle.

This cut was necessary so that when the spoke was driven into the hub it would not project into the space which would be occupied by the bearing. Otherwise it would touch the bearing, and this would cause the spoke to work loose in the hub or split. This cut was made at an angle because the spoke would be inserted in the hub at an angle to create the dish of the wheel.

The dish of the wheel made another cut necessary. The hind side of the foot of the spoke had to be shaped so that it would allow the spoke to assume the angle that created the dish. This cut was made on the table shaper as shown in Figure 70. The spoke was placed in the jig, and as the jig was drawn across the cutting head it cut the desired shape on the foot.

The final step in the preparation of the foot of the spoke was the cutting of a chamfer on the flat side of the foot and shoulder of the tenon. This was done so that the spoke would not split as it was driven into the mortise of the hub. Figure 71 shows the various steps in preparing the foot of the spoke.

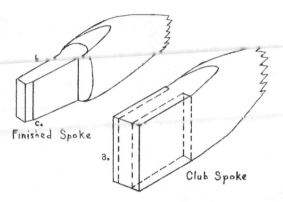

Figure 71. Preparing the foot of the spoke: (a) indicates the tenon cut, (b) is the angle cut onto the hind side of the foot, (c) are the chamfers cut on the foot and shoulder of the tenon.

When spokes were made entirely by the hand process the chamfers were cut with the drawknife and chisel; however after the shop became busier this process was considered too slow and time consuming. The Grubers then devised a means of cutting the chamfers quickly and efficiently through the use of jigs on the table shaper. Figure 72 shows the jig in use on the table shaper. The spoke was placed against a guide in the sliding tray, and as the cutting head rotated the foot of the spoke was pushed across the table cutting one side. To cut the other side, the spoke was removed, the tray was retrieved, and the spoke was replaced against the other side of the guide with the cut side down. This operation completed the preparation of the foot of the spoke.

Felloes

The felloe was that part of the wheel that formed the rim. It was made by wheelwrights since 2000 B.C. much the same as it was at the time the Grubers made wheels.[29] They were made from ash or oak; however, the Grubers used nothing but oak, because they thought that the ash did not last as long as the oak. Felloes were also made by several different methods. If they were

---

[29] *Ibid.*, p. 24.

Figure 72. Jig in use on the table shaper.

made by steam bending, they were made in one or two pieces. They were also shaped by sawing. In the latter method, the circumference of the wheel was divided into equal segments, each large enough to contain two spoke ends. The Grubers generally used the two-part felloe, which they procured in large lots from commercial sources. The felloe was generally not processed until the wheel was ready to be assembled.

Assembling the Wheel

Wheels were made in the same manner in which other parts of the wagon were made; in sets sufficient to make six to twelve wagons at one time. Sufficient numbers of hubs, spokes, and felloes were brought from their various storage areas in the works, to the bench shop. There the work of assembling the wheel began at the wheel pit as shown in Figure 73.

The wheelwright clamped a hub, which had been previously prepared, in the cradle. A spool was placed in the hole in the hub, and a dish gage was clamped to the face of the hub. An iron pin was set in a hole in the upper part of the gage at about where

127

Figure 73. Wheel pit with hub clamped in place ready to have the spoke driven into place. The wooden pot contained cement, which was used only in replacing older spokes. The bridle stick and sledge hammer are shown in the foreground.

the end of the spoke would be after it was driven into place. This pin was adjusted so that it projected from the gage one half to five eighths of an inch less than the distance between the face of the hub and the end of the mortise hole. This distance provided the dish in the wheel that the Grubers considered best. Figure 73 shows the wheel pit with a hub in place.

The wheelwright took his place in front of the pit with a sledge hammer in hand. The helper straddled the pit on the right side with the bridle stick in hand. The wheelwright drove the spoke into the mortise while the helper guided it with the bridle stick so it lined up with the end of the gage pin. Each spoke in turn was driven into the hub and lined up as it was driven.

After all the spokes were in place the wheel was taken to the wood working room where the wheel was placed in the spoke tenoning machine. The spokes were all cut to the same length, then each spoke had a round tenon cut on the head of the spoke.

The wheel was then returned to the bench room where the wheelwright placed the wheel in a special clamp attached to the top of his bench. The wheel was placed hind side down and clamped through the hole in the hub. A felloe was laid on top of the spokes against the shoulder of the tenon in the position where it would be when finally assembled. The center of each spoke was marked on the inside of the felloe, and a hole the size of the round tenon of the spoke was bored on each mark in the center of the felloe.

Before the felloes were put on the spokes each spoke had a saw kerf cut into the head. This cut was made across the spoke so that the pressure of the wedge, to be driven into the kerf later, would not split the felloe.

The spokes were inserted in the holes in the felloes one at a time, each having to be forced into line with a spoke dog, because the distance between the spokes at the head was greater than at the shoulder of the tenon, at which point the holes were spaced. After all spokes were started the felloes were slowly tapped into place, and as the ends of the felloes came together they were trimmed with a saw to make them fit. Great care had to be taken at this point to insure that the wheel would not be felloe bound. This condition existed when the felloes were tightly fitted together and the shoulders of the spokes were not properly seated on the felloes.

After the felloes were properly seated on the spokes the wheelwright drilled a hole into the outside of the felloe that passed through the joint of the felloes into the adjacent felloe. Into this one-half-inch hole a dowel was driven to secure the joint. Wedges made from spoke ends that had been cut off previously were driven into the saw kerfs in the heads of the spokes to insure that those joints would be tight, and the projections were cut off with a saw.

The wheel was now returned to the spoke tenoning machine, which was now fitted with a shaving head. As the wheel was rotated the rotating blades trimmed the outside of the felloes to a true diameter. The wheel was now ready to have the iron tire placed on it.

Tiring the Wheel

Putting the iron tire on the wheel, or "tiring the wheel" as the wheelwright called the operation, required all the skills and knowledge that the blacksmith possessed. If it was not done properly the tire would either be loose or shrink too tightly and it would crush the wheel.

In preparation for this job the blacksmith laid out the long pieces of iron bars on the floor of the shop. A rough measurement was made by rolling the wheel on the bar. Starting by placing one of the felloe joints at one end of the bar, the blacksmith rolled the wheel along until that joint again came to rest on the iron. After adding two and one half inches to this measurement, the bar was cut in the powerful power shear. The short lengths of the bars left from this cut were welded in the forge and used for another tire.

The end of the bar that had been cut for the tire was heated in the forge and bent slightly on the horn of the anvil so that it could be fed into the tire bender or power form roller. The bar was formed into the approximate diameter of the tire to be made. At this time the wheel was measured accurately with a traveler, a small metal wheel that had graduations around its circumference. This measurement was transferred to the roughly shaped iron ring, which was now trimmed to that dimension plus one half inch. When this cut was made the ends of the iron ring were cut in preparation for the welding of the ends. Many blacksmiths made this weld by scarfing the ends of the metal as Kauffman describes it.[30] The Grubers, however, had their own method of making this weld. They cut the ends of the bar on a forty-five degree angle as illustrated in Figure 74. The ends were then positioned so that the short side of the cut overlapped by eleven-sixteenths of an inch. This made the tire one half inch less in circumference than the wheel and was considered satisfactory for the shrinking process that held the tire on the wheel. The ends of the tire were heated to a welding temperature in the forge

---

[30]Henry J. Kauffman, *Early American Ironware*. Rutland, Vermont: Charles E. Tuttle Company, 1966. p. 126.

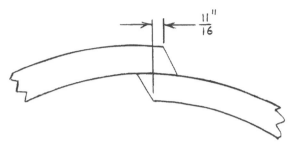

Figure 74. The ends of the iron bar cut and in position for welding.

and welded on the power hammer. With a little hammer work on the anvil this weld could not be seen.

Before they acquired the tire setter, the Grubers put tires on wheels by heating the tire. The wheel was placed face down on a three legged stool as shown in Figure 75.

The tire was heated by several methods. Between 1884 and 1898 a fire was built on the ground. Several tires were piled on the ground, and a fire was built over them. When they were

Figure 75. Three legged stool used to hold the wheel as the hot tire was being put on.

brought to the proper heat they were carried by tongs to the wheel and layed in place. Water, carried in buckets from the dam across the road, was poured over the tires to cool them and shrink them into place. In 1898 a brick oven was built to heat the tires. The tires stood upright in this oven; however, when they were removed they were handled as before.

In 1908 the Grubers changed their process of putting tires on wheels. They purchased a hydraulic tire setter, which improved the process and was much quicker. The process was better because the tire was set cold, and it was done with greater accuracy. It saved time and labor because it did not require the heating process, and it was done in a matter of minutes. The Grubers advertised that they could set a tire while the customer waited, which was not possible with the heat-shrinking method.

In preparing the tire to be put on the wheel by this tire setter the cutting and welding was done in the same manner as before; however, the tire was made so that the diameter was one half to five eighths of an inch larger than the diameter of the wheel. The tire setter was prepared by placing shoes of the proper size in the machine. The wheel was laid in on the bed of the setter, face down, with one-eighth-inch shims under the felloes. This was done so that the tire, which was one quarter inch wider than the felloes, would be centered on them when compressed into place. The tire was slipped over the outside of the wheel and laid on the base. The hydraulic pump was started, and as soon as the pressure was built up, the control valve was opened, and the pressure was applied to the shoes, which closed against the tire and compressed the tire onto the wheel. Since there was no need to heat the tire in this operation, it saved time, did a cleaner job, and was consistently more accurate.

After removing the wheel from the tire setter the blacksmith took the wheel back to the anvil or the power press and put the inside and outside hub bands on. The wheel and these hub bands were all secured with pins driven through holes that the blacksmith drilled through the metal into the wood. Two holes were drilled into the hub bands and one quarter-inch hole was drilled through the tire between every other spoke. The pins were driven into the holes, peened down, and the wheel was ready to have the hub bored.

This operation, as many others, had been done by hand for many years. However, as the Grubers mechanized other processes, they did likewise with this one. The wheel was mounted with the front side to the frame and the hub clamped in the vise. The boring tools designed to shape the hole for the bearing were alternately run into the hub half way. Then, without removing the wheel, the boring tools designed to shape the hub to receive the collar and retaining nut were placed on the machine, and they were used.

The wheel was now returned to the blacksmith shop to have the bearing pressed into place. With the hub bands in place the hub was now ready to withstand the extreme pressure exerted on it when the bearing was pressed into place.

The wheel was placed in the power press face down with the hub resting on the anvil. The bearing was put in the hole, and the press was brought down to force it into place. This was a very precise operation compared to the method used by the English wheelwrights, who formed the bearing hole with chisels and gouges, then positioned the bearing with wedges to true it up.[31]

After this operation the wheels were taken to the paint shop on the second floor and given one coat of vermillion paint, after which they were placed in storage until needed.

THE CARRIAGE

The design and construction of the carriage, or undercarriage as Jenkins refers to it, had many demands on it. It had to be strong enough to support the weight of the load that it was going to carry, yet light enough so that it would not place too much strain on the horses that were going to draw it. It had to be made in such a manner that it distributed the load evenly to each wheel,

---

[31]Jenkins, *op. cit.*, pp. 80–81.

yet designed in such a manner that all the weight would not shift to any one of the wheels at any time and thereby cause a breakdown. Sturt accomplished this through skill and the judgment of the eye; however, the Grubers, through their experience and ingenuity, developed a method by which they could accomplish this same end.

Through the use of mass production methods, interchangeable parts, and jigs and patterns the Grubers produced a carriage that was renowned for its reliability and performance.

The carriage, like the other parts of the Gruber Farm Wagon, was mass produced. The process began by bringing two or more wagonloads of heavy oak planks to the bench shop. The wood was passed through the window to the left of Jacob Gruber's bench. This placed it in the vicinity of the jointer, which was the first machine used in the process of preparing the lumber for use.

The lumber was inspected, and planks that were three inches thick were selected first. These were run through the jointer, and both edges were surfaced. Next they were run through the planer, and both sides were surfaced to a thickness of two and three quarter inches. Patterns of the hounds, bolsters, bolster beds, and axle beds were laid on these planks, and while the gugs or pins, which were strategically located in the pattern, marked the location of centers, positions of dado cuts, and centers of holes, the outline of the pattern was traced on the plank.

This process was repeated until all of the three-inch planks were used. These two-and-three-quarter-inch-thick parts were for use on the two-and-one-quarter-inch axle, two-and-one-eighth-inch axle, two-inch axle, and the one-and-seven-eighth-inch axle wagons.

These shapes were next cut out on the band saw. The cut was made about one eighth inch oversized to allow for shaping. After the parts were cut out on the band saw the pattern was replaced on the part. The gugs in the pattern were replaced in the holes they had made when first put on the plank. This positioned the pattern exactly as it had been before.

The rough-cut part with the pattern in place was run through the curved surface shaper, and through the use of the pattern the exact shape of the part was formed. The parts were then taken to the table shaper on which the square corners on the curved side of the parts were chamfered. After this the parts were returned to the bench shop, the dadoes were cut on the dado machine, and the holes were drilled where required on the boring machine. These machines did such an effective job that no sand paper was needed to smooth the surface of any piece; however, before the shapers were acquired, the belt sanders were used to remove the saw marks of the band saw.

As soon as these parts were finished they were taken to the paint shop, given one coat of vermillion paint, and taken to the storage area and put away until needed.

The small pieces that were left from this operation were used for the smaller parts. Sliding bars, body rockers and supports were made from two-inch stock while the rest was planed down to a thickness of one and one half inches for general use.

Through this mass production method the Grubers made as many as twelve sets of parts for the various sizes of wagons, and, although they did not measure one part at the time it was fabricated, they were confident that the part would fit interchangeably with other component parts at the time of assembly.

THE BODY

The body of the Gruber Farm Wagon was a well proportioned box-like structure. It had none of the curves of the English farm wagons that Jenkins describes.[32] Even so, this body was capable of carrying loads up to six tons if made large enough to contain them. The body was made of poplar boards with oak supports and braces by the same mass production methods as the rest of the wagon.

The Box Body

In preparation for making bodies several wagonloads of

---

[32] *Ibid.*, pp. 15–17.

poplar boards were brought from the lumber storage shed and passed through the window as the lumber for the carriage was. Wagon bodies were not made for storage; they were constructed as needed when the wagons were being assembled. This, however, was still a mass production, for wagons were generally assembled in groups of twelve. Each size of wagon had a different sized body, as shown in Table 3. Boards that were suitable for the largest sizes were selected first and joined on one side. They were then planed to one and one half inches in thickness, which prepared them to be cut on the table saw.

Every piece of lumber that was cut for the body was cut through the use of jigs. They were used to cut the bottom boards of the body, as well as the sides and ends. A long tray-like affair was used for the crosscuts. It had rollers on the ends to support its weight. Movable pins were positioned throughout the length of the tray to adjust the jig for the various lengths to be cut. When not in use this jig was hung on nails from the rafters over the saw.

After sufficient boards were cut to make twelve bodies, the boards were stacked, and the assembly of the body began on

Table 3. Standard Sizes of the Gruber Farm Wagon Body.

Axle Size	Body Size		
	Length	Width	Height
2 1/4"	12' 1"	3' 6"	21" plus 10"
2 1/8"	12' 1"	3' 6"	21" plus 10"
2"	12' 1"	3' 6"	21" plus 10"
1 7/8"	*		18" plus 8"
1 3/4"	10' 1"	3' 5 1/2"	16" plus 6"
1 5/8"	10' 1"	3' 5"	15" plus 6"
1 1/2"	10' 1"	3' 5"	14" plus 5"
1 3/8"	8' 1"	3' 5"	14"
1 1/4"	8' 1"	3' 4"	13" or 14"

*This dimension could be either 12' 1" or 10' 1".

Note: The height of the body was made in two parts. The smaller or top section could be removed to lower the sides of the body if necessary.

horses set up in the bench shop. The proper number of boards for the bottom were joined, then laid on the horses and clamped in wood clamps. Cross pieces were placed at the forward and rear end of the bottom and nailed to the boards. One or two body rockers were then attached to the boards. If only one was to be used, it was placed at about the middle of the body. These pieces were two inches wide, two inches thick and projected four inches on either side of the body. This piece was sometimes attached at the rear end of the bottom. These pieces held the bottom of the body together. After they were fastened, the bottom was turned over and laid on the horses right side up, and the sides were set in place and held there by clamps. The ends of the body were also put in place at this time. They were positioned between two, one-inch by one-inch pieces of oak that were nailed to the sides. Sometimes the rear end of the body was hinged so as to drop when unpinned; however, it was normally slid in place between these two strips and could be easily removed.

Ironing the Body

The foot rest and seat were cut and fitted together and placed in the body, which was then sent to the blacksmith shop to be "ironed." This process completed the construction of the body by fastening it together with bolts, placing iron straps at strategic points for strength, placing iron plates at points that had excessive wear, and mounting the brake mechanism.

The sides, each of which consisted of one wide board, were held totally to the bottom of the body with "L" shaped iron bands placed at the front, rear, and several points between. Figure 76 shows several of them.

"L" shaped pads of iron plate were placed where the body rested on the bolsters and held there by side braces. A similar wearing plate was fastened at the point where the tires of the front wheels wore the body when the wheels were fully locked. These plates can also be seen in Figure 76. Braces of one-half-inch iron rod also braced the side of the body on the forward and rear bolsters as well as the rockers, as shown in Figure 76. Iron rods that ran from side to side added further strength to the

Figure 76. Left front view of body showing various iron fastenings and other attachments.

body at the front and rear. All iron was bolted to the body with wagon bolts.

Most of the braking mechanism was mounted on the carriage; however, a side brake handle and rachet were mounted on the right side of the body within easy reach of the driver. The foot rest of the driver was mounted with iron brackets and braces. This, and the iron sockets that retained the wooden retaining pins that held the body extension in place, completed the ironing of the body.

Ironing the Carriage

To begin the process of ironing the carriage, the blacksmith and his helpers laid the parts together in the manner in which they would be assembled, on boxes or horses, in the blacksmith shop as shown in Figure 77. They then placed the various bolts and pins, which held the parts together, into the predrilled holes. If the holes were too small they would heat the bolt or pin in the forge, then run them into the hole. Various metal braces that added strength to the carriage were also bolted in place. At points

Figure 77. Carriage parts laid out in preparation for assembly by the blacksmith. The hind hounds are not in the proper position. They will be on top of the coupling pole when assembled.

where there was much pressure and friction, such as the point where the forward bolster rested and pivoted on the bolster bed, a wearing plate was placed. These were precut and kept in stock in the shop. Table 4 gives the dimensions of these plates.

The axles were also fitted and attached to the axle bed at this time. Axles were bought as quarter axles. These consisted of two pieces of square iron, each with a spindle formed on one end and the collar, bearing, washer, and nut in place. Figure 78 shows two such axles. The axles were fitted to the curve of the axle bed, cut and welded together on the power hammer. The older design of this axle had less curve to it than the newer design. After the axle had been welded but before it was placed on the axle bed the blacksmith used a pattern, consisting of a long board with pins placed at specific points in it, to set the ends of the axle to the proper angle. These angles were necessary so that the load of the wagon would be properly distributed through the wheel, and the wheels would track properly. As Jope points

Table 4. Bearing Plate Sizes for Various Sized Wagons.

Axle Size	Plate Size
2 1/4"	14" x 4"
2 1/8"	13" x 3 1/2"
2"	13" x 3 1/2"
1 7/8"	12" x 3 1/2"
1 3/4"	12" x 3 1/4"
1 5/8"	10" x 3"
1 1/2"	10" x 2 3/4"
1 3/8"	10" x 2 1/2"
1 1/4"	10" x 2 1/2"

out, the dished wheel was in use from the fifteenth century on-wards.[33] However, there is no evidence recorded on how the wheel was mounted at that time. Sturt, in discussing the skills of the early English wheelwright, states that the dish of the wheel demanded the downward set of the axle end to transmit the weight of the load through the wheel properly. This in turn caused the wheel to pull away from the body. The axle was bent forward slightly to counterbalance this and draw the wheel back toward the center of the wagon.[34] The jig that the Grubers used achieved this delicate balance with consistent accuracy.

After the axle was prepared in this manner it was fitted to the axle bed. Sometimes the drawknife had to be used to get an accurate fit. Four long bolts held the front axle, axle bed hounds, and bolster bed together. The axle was further held to the axle bed at the outer end by the use of "U" shaped clips. Figure 79 shows one of these clips.

The hounds on the forward carriage were braced to the axle bed with curved iron rods, and the forward ends of the hounds

[33]E. M. Jope, "Vehicles and Harness." *A History of Technology.* ed. Charles Singer, E. J. Holmyard, A. R. Hall, and Trevor I. Williams. London: Oxford University Press, 1956, p. 552.

[34]Sturt, *op. cit.,* pp. 135–137.

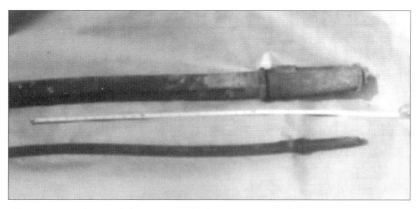

Figure 78. Axles with bearings. The larger is a 2 1/4" size and the smaller is a 1 1/4" size.

were bolted together with iron plates to form a socket for the tongue, which would be inserted later.

The hind axle was assembled in the same way as the forward axle. Then the axle, axle bed, and hind bolster were bolted together with two bolts through them all. Two clips held the

Figure 79. Clips hold the axle to the axle bed at the outer end.

bolster, hounds, axle, and bed together. The axle was further clamped to the axle bed by four clips: two at the outer ends and two on either side of the coupling pole socket.

Various means have been used to stop, slow or prevent wagons from backing down hills. Jenkins describes the use and construction of roller scotches, small cylinders dragged behind the hind wheels when going uphill to prevent their rolling back down in case the horses failed to hold the wagon; dog sticks, which were used for the same purpose and dragged behind the wagon when going uphill; and drag shoes, which were cast iron shoes slipped beneath the front of the hind wheels when going downhill.[35]

The braking system on the Gruber wagon was far more elaborate and effective. It consisted of a brake bar suspended loosely beneath the hind hounds directly in front of the hind wheels. On the ends of this bar directly in front of each wheel were mounted two shoes with wooden liners. From the center of this bar leading back through the hind axle bed was an iron bar that connected the brake bar to a brake handle through a cam-like linkage. Curved back from the hind ends of the hounds and bolted to each end, was a saw-toothed ratchet in which the brake lever, projecting to the rear of the wagon, could be engaged. This linkage could also be operated from the brake lever mounted on the right side of the body next to the driver's seat. The driver's lever was used to hold the wagon in place when it was stopped, whereas the rear lever was used to place a drag on the wagon when going downhill. Figures 80 and 81 show the brake mechanism.

The hind carriage was completed by clamping the forward ends of the hounds together with an iron strap and placing an iron plate on top of the length of the hind bolster to reduce wear between the bolster and body at that point.

The two carriages were now ready to be coupled together. This was done differently than on the English wagons described

---

[35]Jenkins, *op. cit.*, pp. 99–100.

Figure 80. Rear view of the brake mechanism on hind carriage.

Figure 81. Front view of brake mechanism on hind carriage.

by Jenkins. On the English wagons the coupling pole was fitted to the pole socket in the hind carriage, bolted to the forward end of the hind hounds, and pinned to the forward carriage by the king pin.[36] This made a rather rigid connection and did not allow for twist in the carriages as the wagon traveled over the rough roads of that day.

The Gruber wagon was coupled to allow for twist in the carriages. This was done by passing the coupling pole through the pole socket in the hind carriage, then under the forward end of the hind hounds where it was loosely pinned, and finally over the sliding bar of the forward carriage and into the pole socket of the forward carriage, where it was pinned by the head pin. The coupling pole was loosely chained to the head of the hind hounds and by a chain to either side of the forward bolster. This completed the construction of the carriage, and it was now ready to receive the body.

Many wheelwrights held the body of the wagon in place with pins. Sturt used a one-and-one-quarter-inch pin through the body and into the forward bolster and a five-eighths-inch pin through the body into the hind bolster to hold the body in place.[37] The body of the Gruber wagon rested on the bolsters without any fastenings to hold it down. To prevent it from moving forward or backward a strip of oak was bolted to the bottom of the body so that it was against the after edge of the forward bolster. To prevent the body from moving back, a like strip was bolted to the bottom of the body so that it was against the forward side of the rear bolster. The body was prevented from moving from side to side by standards, which were shaped boards fitted and bolted into mortises cut into the ends of the bolsters at the sides of the body. The feet of the standards were tapered on the outside edge so that they would wedge the body tightly as they were pressed into the mortise with their edge against the sides of the body.

---

[36] *Ibid.*, p. 86.

[37] Sturt, *op. cit.*, p. 68.

A tongue or shafts, prefabricated as the other parts were, purchased from some commercial source, were selected from stock and fitted to the wagon depending on whether it was a one-horse or two-or-more-horse wagon. Pole-chains, and stay-chains, which were for the most part purchased, were fitted also according to the manner of draft, as were single trees and double trees. With the fitting of these items the wagon was completed and ready to be taken to the finishing room and painted.

Painting the Wagon

The manner in which the paint was applied, its color, and decorative effects were as much of a factor of identification as the construction and shape of the wagon. In England, wagons were readily associated with geographic regions by the color of their paint and the decorative touches of the craftsmen who applied them.[38] This was also true of the Gruber Farm Wagon.

The Gruber wagon was known to all the local farmers by its colors as well as its design. The carriage was painted Venetian red and the body was green. After the wagon was in the paint shop the body was laid upside down on horses and given one coat of green paint mixed with linseed oil, and a second coat of green mixed with turpentine. The bottom was given a coat of varnish and left to dry until the carriage was painted.

The carriage, constructed of parts that already had one primer coat of paint, received a second coat of Venetian red mixed with linseed oil, and a third coat of Venetian red mixed with turpentine. After this last coat dried the painters applied their decorative touch to the carriage. Every part of the carriage was touched by the thin lining brush of the painter to create thin long curved lines, scrolls, and petal-like designs. Nothing escaped this touch; even the brake shoe was decorated. The carriage was given a coat of varnish, and this completed the finish on it. The body was then placed on the carriage, and it was also decorated with lines and scrolls. The words "Gruber Wagon" were lettered on the sides and back of the body, while on the back of the hind

---

[38]Jenkins, *op. cit.*, pp. 104–105.

bolster the words "Gruber Wagon Works, Mt. Pleasant, Berks Co., Pa." were lettered. The inside of the body was given one coat of Venetian red, and the outside of the body was given one coat of varnish to complete the finishing process.

Serial Numbers

Although the practice varied over the years, the Grubers identified their wagons by serial numbers. Each body as it was completed was stamped on the head boards with four numerals; likewise, each carriage was stamped on the rear of the hind bolster and the forward hound with a number. These numbers were recorded in a ledger, and the Grubers were able to tell who purchased each carriage and body and when it was purchased.

By standardizing their product, mechanizing their processes, and using mass production methods, the Grubers were able to make as many as 75 wagons during the winter months and thereby meet the demands for their product when the farmers needed them in the spring of the year.

CHAPTER VI

THE GRUBER FARM WAGON

The Gruber Farm Wagon consisted of a well proportioned box body placed on a simple but strong carriage with well balanced wheels. It had none of the curves and bulk of the English box wagon described by Jenkins.[39] It was smaller in size yet able to carry heavier loads, as is evidenced by the fact that the English box wagon was designed to be drawn by one or two horses, and the Gruber wagon was made in sizes from one horse to six horses capable of carrying loads from 1 1/4 tons to 6 tons as shown in Table 5. The craftsmanship exemplified in the construction and the use of materials was superior to that shown in the European wagons shown by Jenkins.[40] This superior construction was evidence to support the fact that American technology was more advanced than European technology of that era. It was the application of this advanced technology, and the innovations it brought about, that made the Gruber Farm Wagon a superior product.

Wheels

The Grubers used the same basic methods and construction that wheelwrights had used for centuries. One of the innovations applied by the Grubers in the construction of wheels was the Rousse band as shown in Figure 82. This band projected out over the end of the hub, and many farmers felt that it preserved the wheel bearing because it prevented dirt from getting into the bearing. It could be applied to most any wooden hub. The Grubers normally used the simple iron bands.

---

[39] *Ibid.*, pp. 114–178.

[40] *Ibid.*, pp. 46–47.

Table 5. Dimensions and Data on the Gruber Farm Wagon.

Axle Size	Wheel Diameter		Tire		Spoke Width	Hub	
	Front	Hind	Thick.	Width		Diameter	Length
2 1/4"	3' 2"	3' 10"	7/8"	3"	2 3/4"	9 7/8"	11"
2 1/8"	3' 2"	3' 10"	7/8"	3"	2 1/2"	9 3/8"	10"
2"	3' 2"	3' 10"(1)	3/4"	3"	2 1/2"	8 7/8"	10"
1 7/8"	3' 2"	3' 10"	3/4"	3"	2 1/4"	8 1/4"	9 1/2"
1 3/4"	3'	3' 6"	5/8"	2–3"	2"	7 3/4"	9"
1 5/8"	(2)	(2)	(2)	(2)	1 3/4"	7 1/8"	8 1/2"
1 1/2"					1 1/2"	6 5/8"	8"
1 3/8"					1 3/8"	5 7/8"	7 1/2"
1 1/4"					1 3/8"	5 1/4"	7"

Axle Size	Carriage Thick. of Parts	Body				Horses	Tonnage
		Length	Width	Ht.	Ext.		
2 1/4"	2 3/4"	12' 1"	3' 6"	21"	10"	6	6
2 1/8"	2 3/4"	12' 1"	3' 6"	21"	10"	4	5
2"	2 3/4"	12' 1"	3' 6"	21"	10"	4	3
1 7/8"	2 3/4"	(3)		18"	8"	4	2
1 3/4"	2 1/2"	10' 1"	3' 5 1/2"	16"	8"	2	2
1 5/8"	2 1/2"	10' 1"	3' 5"	15"	8"	2	1 1/2
1 1/2"	2 1/4"	10' 1"	3' 5"	14"	5"	1	1
1 3/8"	2 1/4"	8' 1"	3' 5"	14"	5"	1	1
1 1/4"	2 1/4"	8'	3' 4"	13"	5"	1	1

Note (1): this dimension was later changed to 3' and 3' 8".
Note (2): dimensions varied below this size.
Note (3): the dimensions of this body varied, either 12' 1" or 10' 1".

Another innovation occasionally applied by the Grubers was the patented hub. There were two on the market; the Sarven Patented hub, which was two parts bolted together, and the Warner Patent, which was a cast iron casing into which a wooden cylinder was inserted to hold the bearing. Figure 83 shows a wheel with the Warner hub.

Figure 82. Wheel hub with Rousse band installed.

Figure 83. Wheel with Warner Patented hub.

Figure 84. A wooden axle with a metal skein mounted in place.

Axles

The iron axle was used on wagons as early as 1840 according to Jenkins.[41] The Grubers used a wooden axle as late as the 1890's when requested by the customer. Their axle was not as crude as the axle described by Sturt.[42] The Gruber axle was carefully designed and accurately made as indicated in Figure 84. On the ends where the wheel was fitted a cast iron fitting called a skein was carefully placed. Two types were used: a second type is shown in Figure 85. The skein in Figure 85 has a bearing in place, while the one in Figure 84 has none.

Carriage

The carriage of the Gruber wagon was well constructed and strongly and carefully fitted together. In addition to having the same basic parts as the European wagon it had an effective brake system. A definite mark of identification was the spools on the carriage used to cover the long bolts that ran through the bolsters and into the axle beds to tie them together. These small barrel-shaped pieces of oak were fitted into the space between the bolster and axle bed. Besides covering the bolt, they also acted as a spacer, insured that the bolster and axle bed were properly spaced, and that the reach pole would not fit too tightly in the socket between the two. Figure 86 shows such a spool in the hind carriage.

---

[41] *Ibid.,* p. 83.

[42] Sturt, *op. cit.,* p. 137.

Figure 85. Metal skein used on wooden axles.

Slip Tongue

The tongue or draft pole was made of oak and was any-where from 8 to 12 feet long. It was bolted onto the forward hounds and generally prevented the wagon from being stored very easily. To remedy this the Grubers devised a means of eas-ily taking the pole apart for storage. This innovation was called a slip tongue. A sample of one is shown in Figure 87. The tongue was uncoupled merely by removing the pin from the hinge joint in the tongue.

~

Unless one was very familiar with the Gruber wagon it was difficult to tell the difference between the various sizes. Figures 88, 89, and 90 show three different sized wagons, each looking like the other; however, each one is different. The Gruber Farm Wagon was a well designed and efficiently manufactured prod-uct of early American industry.

Figure 86. Hind carriage showing a spool mounted between the bolster and axle bed.

Figure 87. Slip tongue designed by the Grubers.

Figure 88. Gruber Farm Wagon, 1 1/4 inch axle for one or two horses, capable of carrying 1 1/4 to 1 1/2 tons.

Figure 89. Gruber Farm Wagon, 1 5/8 inch axle for two horses, capable of carrying 1 1/2 tons.

Figure 90. Gruber Farm Wagon, 1 3/4 inch axle for two horses, capable of carrying 2 tons.

CHAPTER VII

PRODUCTS OF THE GRUBER WAGON WORKS

Although the Gruber Wagon Works was primarily known for the fine farm wagon that it produced, it was also capable of producing many other things. Among its products was a variety of wagons, farm implements, and tools. Later, as the automobile came into existence, it sought to keep pace through the construction of truck bodies. Besides the manufacture of these products, the Grubers also repaired many items.

Wagons

Besides the standard farm wagon the Grubers made a hay bed that was compatible with the carriage of the farm wagon. It was made either sixteen or eighteen feet long and six feet four inches wide at the top of the bed, while it was only four feet two inches wide at the bottom. The oak framework was covered with beaded boards generally made from yellow pine or some other soft wood. The length of the body and the distance between the bolsters required that it be strengthened with iron rods on either side, leading down at an angle from the side of the body, under the cross-braces beneath the body and tightened with a turn-buckle in the center. This body was painted the same color as the box body and the carriage was painted Venetian red. To permit the bed to carry a greater load of hay it was provided with ladders, which extended above the body at the front and rear. These ladders were constructed much like a regular ladder except that they were narrow at the top. These ladders were easily installed and removed. Figure 91 shows the details of the hay bed. The small tool box which can be seen mounted on the head board

Figure 91. A Gruber hay bed. This bed is eighteen feet long.

was put on all hay beds built in the Gruber Wagon Works. The body was generally placed on a one and three-quarter-inch axle, although the older beds were put on a two-inch axle. The only difference in this carriage was the longer coupling pole, which was necessary due to length of the body. This wagon was capable of making a very sharp turn because the front wheels, when turning, would fit beneath the body of the wagon. Wheels that were capable of turning like this were known as "full locking" wheels; this resolved a problem that constantly plagued wheelwrights, because insufficient lock restricted the turning capability of the wagon.

Drag Wagon

The Grubers made many forms of wagons to the specific order of their customers. One of the unique wagons made in the history of the works was a drag wagon made to the order of H. Kalbach and Son of Bernville, Berks County:

H. Kalbach & Son
Dealer in
Flour, Feed, Grain, Lumber and C.,
Bernville_Pa_

Mr. Franklin Gruber
Heister's Mills

Dear Sir: Enclosed Please find the following data
for drag wagon wheels.

Diameter Wheels	34 in.
Diameter Nave	12 in.
Dimension Fellies	3x3 in.
Bore, Nave	6 in.
Dimension, Axle	6x8 in.

Finish wheels at your earliest convenience, we saw
Bal. stuff and send down iron

Yours with Pleasure
H. Kalbach & Son

The Grubers were constantly making things to the specific order of their customers, and this was just one of the items. This wagon as it was called, although it was in reality a two wheeled cart without a body, was used by the lumber men to haul the logs out of the forests from which they were cut. Since there were no roads through these areas and the terrain was rough, this was an ideal vehicle for the purpose. One end of the log was raised and chained to the heavy axle, and the log was dragged to the sawmill where it was sawed into planks and boards. Although Mercer describes several vehicles used in Bucks County to haul logs from the woods to the mill, he makes no mention of a vehicle of this type, so it might very well have been a local design.[43]

Special Wagons

Many other wagons were ordered to the specific needs of the customer. One example of such an order is given in the following letter:

---

[43]Mercer, *op. cit.*, pp. 38–41.

N. F. Hartman
North Star Cigar Factory
Adamstown, Lanc., Co., Pa. April 22, 1892

Mr. Gruber
Dear Sir:

Have you any Four Horse Wagons on hand I
wouldt like to have a wagon for use on the street
I do the Hauling for the Adamstown Hat Factory
and want a wagon to carry about 2 or 2 1/2 tons a little
high on the wheels & not to broad
Please let me know the price of such a wagon
without Body and if you have non on hand how soon
you couldt make one for me I must do something
shortly or get mine repaired.

Respectfully yours
N. F. Hartman

The Grubers made wagons for several ice companies in
Reading at various times. They made such wagons for the Reading Cold Storage Company. The Angelica Ice Company, which
advertised that they had 30,000 tons of ice available for shipment, inquired if the Grubers could build them some wagons in
the following letter:

Angelica Ice Company
26 South Sixth St.
Reading, Pa.   March 12, 1886

Mesrs Gruber Bros. Wagon Builders
Mt. Pleasant Pa.

Gents we want to get 2 Ice Wagons Built, such as we
have, would you please send us the lowest price you
could built them for us and how soon you could do
the work

Yours truly

Angelica Ice Co.
A. Ahrens

Feed and grain dealers, lumber dealers, powder companies, and stone and sand companies all wanted special wagons for special reasons. The Grubers made these wagons to the exact specifications of the customer. One of the heaviest wagons manufactured in the shop was made with four inch axles. The rest of the carriage was proportional in size; however, the body was no larger than usual, only stronger in structure. There was only one wagon built with this size of axle; the rest were all built within the limitations of the two and one-quarter to one and one-quarter-inch axle sizes. Variations, to satisfy the customer's needs were made in the construction of the body, wheels, and brakes, as indicated in the following letter:

East Berkley Mar 19, 1892

Mr. F. H. Gruber

Dear Sir:

I want the wagon with the 3 1/2 in. tire 12 & 14 spokes 3/4 in front and 1/8 iron on hind wheels, the axle proportionately strong with the wheels. I measured my old wagon when I came home and it is 3 ft 8 in on the bolster but, I think that I can fix my body that it will do for 3 1/2 ft. I want a good strong four horse wagon with well seasoned white oak no ash used in its construction. Did I understand you, to say that the front wheel turns under those hay-beds I saw in your shop. I would want it to turn under the body if you can possibly make it so. I want it 18ft long and would you make wagon and bed till the latter parts of May. If I recollect right the wagon is to cost $85 and the hay-bed $35. So if you can make both for $120 you may do so. You to deliver wagon and bed when finished.

If there is anything that I omitted please let me know, also let me know whether my terms are acceptable.

I do not want to cut you in your figure but I would like to have a good strong four horse wagon constructed in a workman like manner.

Yours
Charles Dunkel

P.S. Will have money to pay for wagon as soon as finished.

This wagon normally had a three-inch-wide tire that was seven-eighths-inch thick. Apparently the specifications that the customer made for the thickness of the hind tire were in error, for the Grubers normally did not put that thin a tire on any wagon, let alone one that was this size.

Some customers made very insistent demands in the construction of the wagons they ordered. A letter from a flour and feed dealer is a good example of such a demand:

Edw. S. Wertz
Roller Process Flour Feed & Co.
Reading, Pa. July 28, 1890
F. Gruber Esq.

Dear Sir

Morris Geig says that you can not make the <u>Brake</u> in the front that <u>Seams</u> <u>Strange</u> to me and I am sorry for it compels the driver to be always in the middle and not only that but keeps the wagon <u>Box</u> more level If you only could see Mr. <u>Beaty</u> flour wagon I am confident you could arrange so as to be <u>sadisfactory</u> if you must make it at the side of <u>Wagon</u> under no circumstances dare the driver when <u>oper</u> reach behind the top of <u>Seat</u> and what good will it do to extend the <u>Side</u> <u>Boards</u> to front end and then cut a <u>Slit</u> through and put the <u>Break</u> <u>Lever</u> in I can <u>neve</u> put a <u>Bag</u> on it in that condition and might as well not be there Please do the <u>Best</u> you can and if you get <u>Stuck</u> call me at once.

Resp.
E. S. Wertz

There were not many customers who made such insistent demands; however, the Grubers were able to satisfy their needs in most cases.

Other Products

The skills and ingenuity of the Grubers enabled them to make many products besides wagons. They made various types of farm implements such as plows and harrows. Another product made in large quantities was wheelbarrows. The wheelbarrows were made almost entirely of wood. The only iron used in their construction was that in the axles, tires, and braces. They were painted in the same manner as the wagons. They were given one coat of primer, one coat of Venetian red, decorated and varnished. Figures 92 and 93 show the construction and painting.

The blacksmiths were also diversified. They did the iron work for the farm implements and other things that the shop manufactured. One of the products that was solely a result of the blacksmith's efforts was a socket type of wrench. It was made from octagonal stock. Both ends were heated and upset, then a

Figure 92. A Gruber wheelbarrow.

Figure 93. A sideboard recently repainted.

hole was drilled into the end of the upset ends. The size of the hole was determined by the size of the wrench. The ends were heated again and the drilled out ends were placed over a die that was the shape and size of the wrench head, and the socket was shaped by beating the metal to the shape of the die. One end of the wrench was then bent to a right angle. Figure 94 shows some samples of these wrenches.

A little grinding and the stamping of the Gruber name and size on the wrench completed it. They were sized by numbers from 00, 0, 1, 2, 3, 3 1/2, 4, 5, 5 1/2, and 6, which was a standard of identification at the time.

The blacksmiths in the Gruber shop were also very capable at shoeing horses. The horses were tied to the wall at the left of the forge, and as many as six horses could be worked on at one time.

Automobile and Truck Bodies

With the advent of the automobile the Grubers shifted their efforts in that direction. They made truck bodies for two local

Figure 94. Wrenches forged in the Gruber shop.

firms: Morris Krietz, general haulers, and Maier's Bakery. These bodies were at first constructed of wood; however, some of the later efforts were of metal. The fact that the Grubers made these bodies was advertised in the Motor Year Book of 1922. They were identified as, "Manufacturers of Bodies for Auto Delivery Car, Motor Trucks, Commercial Car Body Wood Panels for Body."[44]

Although the Gruber Wagon Works was capable of producing these many diversified products, their main product remained the farm wagon for which they became famous.

[44] Motor Vehicle Year Book 1922," Ware Bros. Co. 1922: Phila., p. 440.

CHAPTER VIII

THE CRAFTSMEN

Although F. H. Gruber originated the Gruber Wagon Works, much of its success was due to the contributions of the men who worked in the shop. His brothers, his sons, his grandson, and all the other craftsmen who labored there each contributed something to its prosperity.

F. H. Gruber (1835–1898)

Franklin Henry Gruber (Figure 95) was born of descendants from the German Palatinates on a farm near Robesonia, Berks County, on September 14, 1835. In his youth he apprenticed to his cousin John Henry who had his wheelwright shop near St. Daniel's (Corner) Church, Robesonia. In 1860 he entered the wheelwright business with his brothers in Mt. Pleasant. After several years he left the business to return to farming; however, within a short period of time he was again working at the trade for which he was so well qualified.

His small farm shop soon became too small, and in 1883 he built a new shop across the road from his farm. This too rapidly became insufficient for his needs, and in 1884 he turned his farm over to a tenant farmer and moved into a new and larger shop that he built along the Licking Creek near Mt. Pleasant. He and his four sons started making wagons in this shop in the spring of 1885. They worked from 6:00 A.M. till 12 noon, 1:00 P.M. to 4:30 P.M., and frequently went back to work until a job was finished; sometimes until late into the night.

The Sons

F. H. Gruber had six sons, John William, James Milton, Adam Refusal, Jacob Henry, George Pierce, and Levi Franklin. James

Figure 95. Franklin Henry Gruber, wheelwright and craftsman.

(1862–1863) died at an early age and Levi Franklin (1870–1941) studied for the ministry; however, the other four sons learned the wheelwright trade from their father and worked with him in his shop (Figure 96).

John W. Gruber (1860–1934)

John was first a wheelwright; he worked at his bench in the shop with his father and brothers. He was also a capable blacksmith, but because of his business ability he did much of the buying and later, after his father's death, he became the general manager of the business.

Adam Gruber (1864–1903)

Adam knew the wheelwright trade, but he became the painter in the shop. It was his flowing handwriting that set the style for the scroll-like decorations and trimmings on the wagons. He was killed in a runaway horse accident at the Blue Marsh Bridge just a few miles from the wagon works, while returning from a business trip to Reading.

Figure 96. The Gruber brothers, Jacob, John, and George.

Jacob Gruber (1865–1944)

Jacob was probably the most skilled of all the brothers (Figure 97). He was the master wheelwright, and it was he who designed and made most of the machinery that was made in the shop. He was also a watch and clock maker. He would work long hours into the night repairing watches and clocks that people brought to him for repairs.

George Gruber (1867–1941)

George was the blacksmith in the family. It was not that he did not know the wheelwright trade, for he was well qualified in that respect, but he loved to work with iron. He preferred to make his own tools and made many of the hand tools that were used in the shop. He worked for 52 years at the forge before he finally retired.

Franklin P. Gruber (1884–1978)

Franklin P. Gruber was the son of John. He was raised in the wagon works and learned the trade from his father and uncles. He thereby became an expert in all phases of the wheelwright

Figure 97. Jacob Gruber, wheelwright and master craftsman, at work at his bench.

trade. When his father and uncles retired in 1935 he assumed control of the business.

The Workmen

The men who worked in the wagon works (Figure 98) came from the surrounding farms and communities. In the spring, summer, and fall, when the demands of the farmer were great, they worked their farms and were not available to work in the wagon works. During the winter, when the chores of the farm were few, the men were able to work elsewhere. It was at this time that most of the mass production work was done in the works. Parts were cut, wagons were assembled, painted, and stored.

The men came to work at various times due to their chores at home and on the farm, and they were through at 4:30. Later, in the 1920's, work hours were more standard. The men started work at 6:30, worked until 11:30, at which time they ate lunch, and returned to work at 12:30, to work until 5:30, when they quit for the day. The shop employed as many as twenty men at one time during peak production periods.

When the new shop started in production there was no division of labor; however, it was not long before the labors of the working men were divided. The men worked as woodworkers, blacksmiths, or painters. The Grubers felt that this division of labor enabled them to achieve the high production rates that they did.

Woodworkers

The men in this department worked under the supervision of F. H. Gruber at first and later under John and Jacob. They often went into the woods and returned with the lumber they would use and stored it in the drying sheds. It was their job to get lumber from the drying sheds, take it to the shop, and run it through the processes to make the parts for the wagons. After helping to assemble the wagons they took them to the black-smith shop to be "ironed."

Many different men worked in this area. Some of them worked only in the winter, later some worked all year-round. Some of these men were as follows: James Bricker, John Faust, James Kreider, Andrew Oxenreider, George Ulrich, and Henry Wortman. F. H. Gruber, John, Jacob, and Frank each had a bench in the bench shop. Later, after F. H. Gruber passed away, John Faust worked at his bench. The others worked throughout the bench and wood working shops as they were needed.

Blacksmiths

Although Henry L. Fiant was the first blacksmith who worked for F. H. Gruber, he never came to the wagon works to work permanently. He did, however, come to the works and help out on some difficult jobs when the occasion arose. Jim Rothermel was the first blacksmith who worked at the works. It was he

Figure 98. The Gruber brothers and the workmen at the Gruber Wagon Works. First row left to right: Amandon Bender, Tom Bricker, John Gruber, Jacob Gruber, George Gruber, and John Burkey. Second row left to right: Isaac Black, Jim Kreider, Andrew Oxenreider, unknown, and Milton Kreider.

who taught the trade to George Gruber, who later assumed charge of the blacksmith shop. Next to George, Isaac Black worked the longest period of time in the shop. He worked there 19 years and 9 months. Others who worked in the blacksmith shop were: Tom Bricker, Amandon Bender, John Burkey, Benjamin Bickel, James Daniel, Lewis Davis, George Kissling, Martin Koller, Daniel Ohlinger, Jonathon Ohlinger, James Seyfrit, William Schlappich, Frank Stoudt, and James Weber.

It was the job of these men, who at various times worked in the blacksmith shop, to bring the shipment of iron that arrived in carload lots, from the nearest railroad terminal, at Leesport. They were also responsible for the storage of the iron in the storehouse and for getting the material from there when needed. After fastening the carriage and body together with bolts, pins, and straps, and putting the tires, rings, and bearings on the wheels, they took the wagon to the paint shop.

Painters

When the new shop first opened, Adam Gruber was in charge of the paint shop. Other men who worked as painters were: Nathaniel Bare, Charles Hetrick, Peter Holtzman, Milton Kreider, Samuel Kreider, Elmer Schrack, and Elias Phillip.

Before 1905 these men pulled the wagons up into the paint shop on two planks laid from the road to the large door on the south side of the shop. After the elevator was built the job of bringing the wagons into the paint shop was much easier. Besides painting the wagons these men also put the primer coat of red lead on all parts before they were placed in storage.

CHAPTER IX

THE BUILDINGS

Like other growing things, the Gruber Wagon Works was constantly changing throughout its productive years. One of the areas of change that reflected the growth of the shop most was the construction of the buildings.

The Beginning

In 1854 Isaac H. Gruber, a brother of F. H. Gruber, began to operate a wheelwright shop in Mt. Pleasant, Berks County, Pennsylvania. In 1860 Franklin H. Gruber started practicing his trade in the employment of this shop. He remained at this work for ten years after which time he returned to farming.

The Farm Shop

F. H. Gruber's career as a farmer was short-lived for his friends and neighbors soon started to bring their wagons and farm equipment to him for repairs. He started this work in the small shop that was standard on most Berks County farms. His shop was located across the road from the barn and contained a forge for working on iron and the general hand tools necessary for maintenance of farm machinery at that time. To those Mr. Gruber added the tools of his trade as a wheelwright.

In a short time Mr. Gruber started to make wagons and to do this he acquired some machines. These were placed in the barn on the threshing floor, and it was there that he worked for awhile. As the business grew he soon found it necessary to use another building. At that time he erected a two story building next to the blacksmith shop across the road from the barn. It was in these buildings that he taught his sons the trade, and the

Figure 99. A view of the Gruber Wagon Works showing the addition of the two story farm building on the left.

business grew until 1882. At that time Mr. Gruber started to build another structure to accommodate his growing business.

The New Shop

The new shop was located at the intersection of the road to Reading and Heister's Mill Road. The Licking Creek, which flowed by this point, contained a good supply of water. Construction was completed in 1883, but Mr. Gruber and his sons did not begin to use it until April 1, 1884.

The new building was 83 feet long and 28 feet wide. It had three floors and a basement. Floor plans of the building are shown in Appendix A. It was of frame construction with a slate roof. Within the three foot thick foundation walls the earth was excavated to a depth of six feet. In this area the water turbine and the main drive line were located. The first floor was divided into three rooms. It contained a bench shop that was 34 feet long and 28 feet wide, a wood shop that was 23 feet 6 inches long and 28 feet wide, and a blacksmith shop that was 25 feet 6 inches long.

Figure 100. A view of the Gruber Wagon Works showing the structure that housed the elevator.

The second floor consisted of two rooms. One room was the paint shop that was 33 feet long and 28 feet wide and the other room, which was intended for storage, was 50 feet long and 28 feet wide. The third floor beneath the sloping roof was used for storage and ran the entire length of the building.

Addition to the Wood Shop

Shortly after they started operations in 1885 in the new building the Grubers discovered that they needed more room to accommodate more machinery. In response to this need they removed the two story building from the farm and added it to the rear of the wood shop. This addition is shown in Figure 99 to the left of the original structure. This structure added an area of 24 feet 6 inches by 30 feet 6 inches to the floor space of the wood shop and the storage area above it.

Additions to the Blacksmith Shop

After the death of their father in 1898 the four brothers assumed control of the business. Shortly after that, in 1903, brother

Figure 101. Elevator operating mechanism.

Adam was killed in a runaway horse accident at Blue Marsh Bridge. The three remaining brothers assumed control of the business on January 1, 1904, and the next year they decided to expand the capabilities of the blacksmith shop by expanding that area of the building and installing more machinery. At that time they added an area of 26 feet 6 inches to the rear of the shop as shown in Figure 2 of Appendix A. Two other additions were later made to this area. One was made in 1908 when they needed more room to assemble wagons. These additions are also shown in Figure 2 of Appendix A.

The Elevator

It was not uncommon, in the construction of wheelwright shops, to place the paint shop on the second floor. This removed it from the dust and dirt that permeated the air in the lower areas. When the painting area was on the second floor, wagons were moved to that level through the use of a ramp consisting of two planks leading from the ground level to the paint shop. The wagon was drawn up the ramp by the use of a rope or block and

Figure 102. Front of iron shed showing how iron bars were stored in racks.

tackle. Kauffman refers to a shop in Goodville, Lancaster County, that had such a ramp.[45]

In 1905 the Gruber brothers decided to improve this method of moving wagons to and from the second floor. To do this they built an elevator. The elevator was housed in a structure that was attached to the front of the building. It extended above the third level and was capable of moving wagons from the ground level to the second or third level. The structure is shown in Figure 100.

The elevator mechanism was operated by an endless rope which was led over a large wheel, which in turn was geared to a shaft that had a drum on either end. The wires that were attached to the four corners of the elevator platform were wound around these drums, and as the drums rotated the elevator was raised or lowered. A ratchet assembly held the platform at the desired level. Figure 101 shows the elevator operating mechanism. This

[45]Kauffman, *op. cit.*, p. 125.

elevator saved many man-hours of labor, for it took only two men to move a wagon by this means, whereas it previously took five or six men to do the job.

Outbuildings

At various times during the growth of the business there was need for more storage space. To satisfy this demand, the Grubers built small buildings near the main building. One of the first buildings constructed for this purpose was the small shed seen in Figure 100, which was used to store the iron stock. As this material was, for the most part, in lengths of ten to fifteen feet the building was made only twenty feet long. Later when the suppliers started to ship longer lengths this building had to be lengthened to accommodate the stock. The bars and rods were stored in racks built into the shed as shown in Figure 102. There was a small door at the rear of the shed where smaller stock was kept.

Another building, somewhat larger, was built to the north of the iron shed. This building was used to store lumber to dry and wagons that were completed. In 1935 after Frank P. Gruber assumed control of the business he moved this building to a location across the road from the main building where another building similar in size had been built some time before.

Boiler Room and Engine Room

The original power source was installed in the basement; therefore when the steam engine and boiler were installed in 1896 a special room had to be built to house them. These rooms were built adjacent to the south wall of the wood shop, on the ground level, as shown in Figure 1 of Appendix A. The engine room was in the corner next to the main building foundation, and the boiler room was out at the end of the wood shop. Later, when the gas engine replaced the steam engine, it was placed in the engine room, and the boiler room was used for storage of maintenance materials for the engine.

Office

The smallest addition, yet one that had great significance as to the growth of the Gruber Wagon Works, was the office. It was added in 1905 at the time the elevator was installed. The room projected from the southwest corner of the bench shop and was accessible through a door in that corner of the shop. As can be seen in Figure 99 and Figure 2 of Appendix A, it had five walls of which three had windows. The floor space was only ten feet by five feet, yet it was sufficient to conduct the business of the shop and store the simple records that were kept.

If nothing else indicated that the Gruber Wagon Works was continually growing throughout its productive years, the constant construction and changing of the buildings was certainly sufficient evidence of that fact.

CONCLUSION

The Gruber Wagon Works began as a small craftsman's shop in response to the needs and demands of the community that surrounded it. Within a short period of time the demands for its products and services outgrew the capabilities of the shop and F. H. Gruber started searching for new methods and means to meet the demands.

In attempting to resolve his problems, F. H. Gruber turned to the recent innovations of technology. James B. Francis had recently improved Benoit Fourneyron's water turbine; Eli Whitney had long since proven a system of mass production through the use of machinery; Eli Terry had successfully devised a method of using interchangeable parts in the manufacture of clocks; Samuel Slater had introduced unskilled labor to industry; and such men as Thomas Blanchard were continuously devising new applications for machinery. Therefore, these developments greatly influenced F. H. Gruber's plans for his new shop.

The new shop was located at the intersection of two main thoroughfares near a good source of water. The new power source was one of the improved water turbines which made it practical to utilize more and newer machines in the production processes. Likewise the use of patterns made it possible to utilize unskilled labor and interchangeable parts in the manufacture of the product. These innovations, combined with the skills of the Gruber craftsmen, aided the growth of the business.

However, it was more than the skills of the craftsmen that built the Gruber Wagon Works into a successful business. It was their desire to improve their product, their constant effort to keep abreast of the latest trends of industry, their receptive attitude toward technological change, and their ability to apply these changes to their particular needs that transformed the original craftsman's shop into a successful factory complex. In this respect the Gruber Wagon Works was an excellent example of that transitional stage of American industry.

GLOSSARY

Axle Bed:	The heavy timber onto which the iron axle was fastened. Usually of oak.
Bearing:	Sometimes called the box. A cast iron sleeve which was pressed into the hub to reduce the friction between the rotating wheel and the axle.
Body:	That part of the wagon that contained the load.
Bolster:	The heavy timber on which the body rested. Usually of oak.
Bolster Bed:	A heavy timber on which the front bolster rested. Usually of oak.
Bruzz:	A three cornered chisel used by the wheelwright to clean out the square corners of the mortises he cut into the hub.
Carriage:	The understructure of the wagon, which rested on the wheels and supported the body. Four-wheeled wagons had two carriages, one front and the other hind.
Clip:	An iron fastener which was "U" shaped and was used to fasten various parts of the carriage together.
Dish:	The concave shape into which the wheel was formed. The amount of dish varied with the wheelwright. The Gruber wheels were dished between 1/2 and 5/8 inch.
Dog Stick:	A strong wooden pole used as a safety device on English carts and wagons. One end of the pole was fastened to the rear axle bed and the other end was left to drag on the ground when the wagon was proceeding uphill. If the traces broke or the horses faltered, the dog stick dug into the ground as the wagon backed down, thereby preventing it from running away.
Felloe:	A section of the wooden rim of a wheel. Sometimes there were only two parts or even one; however, if the rim was in sections, each felloe was long enough to contain two spokes. For a twelve-spoked wheel there were six parts, for a fourteen-spoked wheel there were seven parts, and for a sixteen-spoked wheel there were eight parts.

Felloe Block: A block of wood from which the felloe was cut.

Felloe Bound: The condition that existed in the rim of the wheel when the felloes were longer than they should have been, and because of that the spokes were not able to fit as tightly as they should have.

Foot of a Spoke: That end of the spoke that was inserted into the hub. It was generally cut in the shape of a square tenon.

Galling Plates: Plates of iron that were fastened at those points of a wagon where two members met, and there was movement between the two. These plates reduced wear at those points.

Hounds: Heavy curved members of the carriage by which the wagon was drawn. There were two sets; one in the forward carriage and one in the hind carriage.

Ladder: An open frame in the form of a ladder that was used to extend the loading capacity of a hay bed.

Lock: The maximum angle that the front wheels of the wagon were able to turn. A full locking wagon was one on which the wheels could turn under the wagon body.

Locking Cleat: An iron plate that was fastened to the body at the point where the tire of the front wheels would strike the body when locked. This cleat reduced wear at that point.

Mandrel: A large cone-shaped device used by the blacksmith to shape circular objects. Its principal use in the wheelwright's shop was to shape hub bands.

Mesh: That part of the hub left between the mortises cut into the hub to receive the spokes.

Scarfing: The act of thinning down the end of a piece of metal prior to welding. This was done to keep the weld from becoming too thick.

Traveller: A disc with a handle that was used to measure the circumference of the wheel and tire.

Upset: To make metal thicker by making it bulge. Generally done by heating the metal then beating it with a hammer.

BIBLIOGRAPHY

Childe, V. Gordon. *Rotary Motion*. Vol. I of *A History of Technology*. Edited by Charles Singer, E. J. Holmyard, and A. R. Hall. London: Oxford University Press, 1954.

Jenkins, J. Geraint. *The English Farm Wagon*. Lingfield, Surrey, England: The Oakwood Press for the University of Reading, 1961. [Second edition, 1992, Newton Abbot, David & Charles.]

Jope, E. M. *Vehicles and Harness*. Vol. II of *A History of Technology*. Edited by Charles Singer, E. J. Holmyard, A. R. Hall and Trevor I. Williams. London: Oxford University Press, 1956.

Kauffman, Henry J. *Early American Ironware*. Rutland, Vermont: Charles E. Tuttle Company, 1966.

Mercer, Henry C. *Ancient Carpenter's Tools*. Doylestown: The Bucks County Historical Society, 1960. (First published, 1929; reprinted at least four times through 1975 by Bucks County Historical Society; reprinted, 2002, by Dover Publications, Mineola, NY.)

"Motor Vehicle Year Book 1922," Philadelphia: Ware Brothers Company, 1922.

Strassman, W. Paul. *Risk and Technological Innovation*. Ithaca: Cornell University Press, 1959.

Sturt, George. *The Wheelwright's Shop*. Cambridge, England: The University Press, 1934. (First published, 1923; reprinted repeatedly, as recently as 1993, with new Foreword by E. P. Thompson, but without plates.)

Usher, Abbott Payson. *A History of Mechanical Inventions*. London: Oxford University Press, 1954.

APPENDIX A

Architectural Drawings of the Gruber Wagon Works at Mt. Pleasant, Berks County, Pennsylvania.

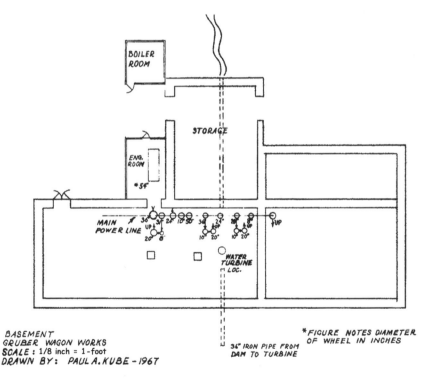

Appendix Figure 1. Basement plan of the Gruber Wagon Works. The foundation walls were stone masonry thirty to thirty-eight inches thick.

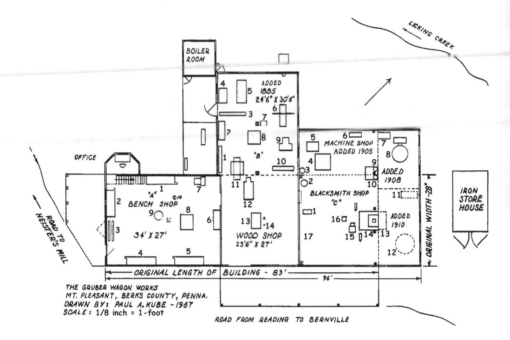

Appendix Figure 2. First floor plans showing the location of machinery and additions made to the original shop. A key to the machinery on this floor follows.

Bench Shop

1	Wheelwright Bench of Franklin Henry Gruber
2	Wheelwright Bench of Franklin P. Gruber
3	Wheel Pit
4	Wheelwright Bench of John W. Gruber
5	Wheelwright Bench of Jacob H. Gruber
6	Jointer manufactured by L. Powers & Co., Philadelphia
7	Boring Machine
8	Dado Cutting Machine made by the Grubers
9	Pot Bellied Stove

Wood Shop

1	Wood Turning Lathe
2	Spoke Tenoning Machine made by the Grubers
3	Belt Sanding Machine made by the Grubers
4	Spoke Tenoning Machine manufactured by H. B. Smith Machine Co., Smithville, N. J. Pat. Jan. 23, 1866

5   Spoke Making Machine manufactured by John Gleason, Philadelphia, 1873

6   Hub Boring Machine made by the Grubers

7   Shaper rebuilt by the Grubers

8   Table Shaper manufactured by L. Powers & Co., Philadelphia

9   Mortising Machine manufactured by Goodell & Waters, 3101 Chestnut St., Philadelphia. Serial #5988

10   Belt Sander

11   Planing Machine

12   Band Saw

13   Table Saw made by the Grubers

14   Control shaft for water turbine

Blacksmith Shop

1   Tire Shrinker manufactured by D. H. Potts, Lancaster

2   Floor Drill Press manufactured by W. E. & J. Barnes Co., Rockford, Illinois

3   Small Floor Drill Press used for reaming

4   Power Punch and Shear manufactured by Royersford Foundry & Machine Co., Royersford, 1894. No. 2 Shop #1438

5   Tire Form Roller

6   Bolt Threader manufactured by Wells Brothers & Co., Greenfield, Mass.

7   Power Press manufactured by Defiance Machine Works, Defiance, Ohio

8   Hydraulic Tire Setter manufactured by West Tire Setter Co., Rochester, N. Y.

9   Bench Drill Press manufactured by Howley & Hermanec, Williamsport

10   Small Bench Press

11   Location of brick tire heating furnace

12   Location where fire was built to heat tires

13   Third and fourth forges built in 1910 and removed in 1925

14   Double forge

15   Power Hammer manufactured by Hawkeye Mfg. Co., Cedar Rapids, Iowa. Pat. Sept. 29, 1903

16   Swage Block

17   Horse Shoeing area for six horses

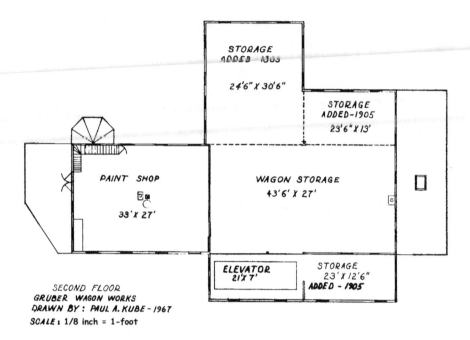

Appendix Figure 3. Second floor plans showing additions.

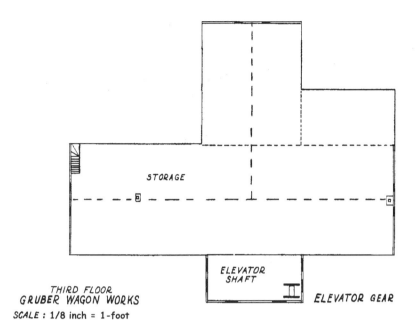

STORAGE

ELEVATOR
SHAFT

ELEVATOR GEAR

THIRD FLOOR
GRUBER WAGON WORKS
SCALE : 1/8 inch = 1-foot

Appendix Figure 4. Third floor plans showing the location of the elevator drive gear.

Legacy and Perspectives

The Gruber Wagon Works Since 1968

Frank P. Gruber (1884–1978), grandson of the founder of the Gruber Wagon Works, took charge of the operation in 1935 and continued to build a few wagons through 1956. After that he kept the shop open with little change, doing wagon repairs until he closed the business in 1971. It was during the last phase of the company's operation that he provided first-hand information to Paul Kube. Most importantly, he kept the Wagon Works and its contents intact, as if the shop had just closed for the night, until the property was purchased by the US Government in 1974. Two years later the building and its contents were moved a short distance downstream, and in 1980 ownership was transferred to Berks County, which assumed responsibility for its preservation, maintenance, and interpretation.

Following is a chronology of the major involvement of the US Government with the Gruber Wagon Works:

1970	Preliminary studies are undertaken by the US Army Corps of Engineers for its Blue Marsh Lake Project, which would inundate the original site of the Wagon Works.
1972	The Wagon Works is listed in the National Register of Historic Places.
1973–1974	The Wagon Works is recorded and documented by the Historic American Engineering Record (HAER) program of the National Park Service.

1974	The Wagon Works is purchased by the Corps of Engineers, beginning the largest relocation/restoration project ever undertaken by the Corps.
1976	In December, the Wagon Works is moved 5 miles from its original site at Mount Pleasant to its present location at the Berks County Heritage Center (Figure VI).
1977	The Wagon Works is declared a National Historic Landmark on December 22.
1978	A historic structure report on the Wagon Works is prepared (John Milner Associates, 1978) and a contract is let for the restoration of the Wagon Works to its condition of 1915, both initiated by the Corps of Engineers. (The Blue Marsh Dam was dedicated June 25, 1978.)
1980	The Wagon Works is deeded by the US Government to Berks County on June 15, following completion of restoration.

Further detail on all phases of the preceding timeline may be found in John Milner Associates (1978), in the HABS/HAER records of the National Park Service, and in the Library of Congress under American Memory. This timeline and its interrelationships to local developments leading to preservation of the Wagon Works as it exists today are discussed in detail by Hunsberger (2005).

The Wagon Works was opened to the public and its 100th anniversary celebrated on May 16, 1982 (Cech, 1982). It has been maintained and operated as a historic site since that time by the Berks County Parks & Recreation Department. A Heritage Celebration is observed every October (attracting some 4000 visitors in 2004), featuring tours of the Wagon Works and demonstrations of

Figure VI. The central section of the Gruber Wagon Works on the road, December 1976, from its original site at Mount Pleasant to its new home at the Berks County Heritage Center. This section, weighing 86 tons, is 71 feet long, 26 feet wide, and 40 feet high. The view is similar to that in Figure II. Photo by US Army Corps of Engineers.

horse-drawn vehicles and related activities (Kunkel, 1989). The Wagon Works is open to the public for a modest fee from May Day to the last Sunday of October. Tours by volunteer guides are offered.

The Wagon Works is located within the Berks County Heritage Center, which also houses historic records relating to the Wagon Works, maintains a collection of its products, and catalogs its contents. The Heritage Center also maintains and improves the structures. For example, funds were sought for the repair and restoration of the deteriorating roof, which effort ultimately resulted in a grant of $195,000, awarded in May 2000, as a part of the Transportation Enhancement Program. This was a cooperative funding arrangement through the Federal Highway Administration and the Pennsylvania Department of Transportation, under the Transportation Equity Act for the 21st Century

(TEA-21). Currently under study, and the most urgent immediate need, is the repair, preservation, and/or replacement of windows and exterior sheathing. Preservation in perpetuity requires constantly keeping an eye upon preventive maintenance. The Berks County Parks & Recreation Department continues to pursue grant funding, both locally and through state and federal programs, to assist in the ongoing care of the building. The Society for the Preservation of the Gruber Wagon Works, established by Frank P. Gruber's daughter Elsie and her husband John, provides financial support for the maintenance and preservation of the Wagon Works. Mildred Gruber Arnold and her husband, Kramer, have dedicated a portion of their estate to the ongoing preservation of the building and the archives.

Now, beyond the established public programs and continuing care of the structures and contents, what of the future? Much needed are further acquisitions of historic records and related research materials, including an active register of location of surviving Gruber products. Perhaps the greatest need is for a suitable building to accommodate additional products of the Wagon Works (some of which are not represented in the collections), along with relevant materials and tools, for storage, display, demonstration, and teaching.

Appendix I

A Biographical Sketch of
Paul Albert Kube

Paul Kube (pronounced Koo´-bee) was born in Chester, Pennsylvania, on 20 December 1918, the only child of Wallace Kube and Sallie Matilda (Berger) Kube. Immediately after graduating from Reading High School in 1937, he entered the United States Navy, where his service included fire-control during battle action in the Pacific Theater of World War II aboard the heavy cruiser USS *New Orleans*. The flag that he salvaged from that ship is on display at the Patriots Point Naval & Maritime Museum, Mount Pleasant, South Carolina. During the 1950s he served in the North Pacific aboard the tender USS *Shenandoah*, where he was in charge of the shops supporting the fleet. After 22 years of service, he retired in 1960 as Chief Warrant Officer third class.

While aboard ship Kube had completed some college correspondence courses, and in 1960 he entered Millersville State College (now Millersville University), Millersville, Pennsylvania, earning his Bachelor of Education degree in 1963 and Master of Education in 1968. He taught industrial arts at Boyertown Area High School (Figure VII) from 1963 until his retirement in 1981. He served in part as departmental chairman and in 1971 was named outstanding electronics teacher by the American Industrial Arts Association. While teaching full time, he completed his master's program under the direction of Professor Henry J. Kauffman, who taught at Millersville University from 1942 to 1973, and had a contagious and knowledgeable enthusiasm for the history, craftsmanship, and industry of southeastern

Figure VII. Paul Kube, while teaching at Boyertown Area High School in 1965. Photo reproduced by courtesy of the *Boyertown Times* (1965:1).

Pennsylvania (see Kauffman, 1999:89–90). Paul Kube loved old barns and initially planned to study the traditional barns of the region as his thesis topic. Fortunately for us, he turned his attention to the Gruber Wagon Works, the subject of our present project. He was preadapted in mindset, ability, and experience to do the unprecedented job that he did on the Wagon Works. His navy duties required a working knowledge of technology ranging from big guns to watch repair. He built a grandfather clock that remains in the family; he was not satisfied merely to weave fabrics — he also built the loom and spun the yarn; he created professional-quality products in copper and pewter. Although he regarded education as the best hope for the World and valuable for its own sake (he tried to learn a new word every day), he was never satisfied with bookish knowledge alone; his commitment to learning included hands-on mastery of the topic, which led to in-depth understanding and appreciation. Thus, he was among the first to work toward preservation of the Wagon Works, which remained of great interest to him for the rest of his life. He died on 7 September 1988, and is buried in the National Cemetery at Gettysburg, Pennsylvania.

Paul Kube was married in 1943 to Marie Rice, with whom he had two children, Paul, Jr., and Sallie. Mrs. Kube died in 1971, and in 1972 he married Sophie Tacyn, who survives him, as do his children.

Production Records of the Gruber Wagon Works

Although highly incomplete, some records of production by the Wagon Works for the years 1904–1935 have been preserved. Table II summarizes the output, by year, of product categories, and tables III and IV present information available, by year, for individual product units.

For our purposes here, these records are especially interesting for the serial numbers of wagon running gears, box bodies, and hay flats that they contain. These numbers, 7/16" in height, generally were stamped into the wood in one or more places on each product. On running gears the number is generally to be found centered on the back of the rear bolster near the Gruber name, on the bolster standards for reception of the box bodies, and sometimes on the chamfered edges of the hounds. On hay flats the serial numbers most often appear on the outside of the framework or on the end grain of the floorboards of the front end, sometimes below the tool box. Wagon boxes were numbered on the front and rear end boards or transoms; wheelbarrows in the center of the top frame board. Although sleighs and truck bodies were numbered, we have yet to find a number on examples seen. The front (rocking) bolsters with standards in place, and isolated standards (or their sawed-off ends) from rear bolsters, turn up at auctions and in antique shops, separated from their running gears, as they had to be removed when hay flats were put in place on running gears, often to remain permanently for all practical purposes, leaving the isolated parts eventually to go astray. The shorter coupling poles suitable for box bodies also appear in shops and auctions, as they had to be replaced by longer poles when switching to hay flats. These longer poles do not show up in isolation, further suggesting that hay flats tended to stay put, once coming to rest on running gears; switching back to box bodies would have been a laborious job, not taken on lightly.

Serial numbers on all products are apt to be obscured by grime and by post-production repainting, but if one knows where the number ought to be, it can generally be brought out by careful cleaning and by close scrutiny in raking light. Failure to find a number does not necessarily mean that the object is not a Gruber product.

We also present here the present location of some surviving Gruber wagons, especially those in public, self-perpetuating repositories where they may be viewed. If any of our readers know the whereabouts of any items not included in our list (especially of Gruber truck bodies, on which we have no current records), that information would be gratefully received at The Berks County Heritage Center, 2201 Tulpehocken Road, Wyomissing, Pennsylvania 19610; e-mail: parks@countyofberks.com; telephone: (610) 374-8839.

Table II. Annual Output of Product Categories by the Gruber Wagon Works, 1904–1934.

PRODUCT CATEGORY

YEAR	RUNNING GEARS	BOXES	HAY FLATS	SLEIGHS	WHEELBARROWS	ICE WAGONS	ICE BOXES	TRUCK BODIES
1904	128	86	44	10				
1905	119	77	51	16				
1906	138	60	67					
1907	? [a]	79	53			10	10	
1908	? [a]	75	47		24			
1909	? [a]	100	61					
1910	182	75	56	?10				
1911	114	98	48					
1912	95	59	53					
1913	138	68	63					
1914	120	81	58					
1915	101	78	69					
1916	98	43	57					
1917	79	55	50					
1918	73	49	68					
1919	91	42	41					
1920	45	57	34					8
1921	36	17	17					18
1922	18	13	12					30
1923	6	11	12					32
1924	9		8					20
1925	14	6	9					11
1926	12	2	9					2
1927	19		28					
1928	21		20					1
1929	6		9					10
1930	7		25					5
1931	8							1
1932								
1933								
1934			7					

[a] During the three-year period 1907 through 1909, a total of 361 units of running gear was produced and given serial numbers.

Table III. Production Records for the Years 1904–1910.

YEAR	BEGINNING AND ENDING SERIAL NUMBERS						
	RUNNING GEARS	BOXES	HAY FLATS	SLEIGHS[1]	WHEEL-BARROWS	ICE WAGONS	ICE BOXES
1904	1 TO 128	1 TO 86	1 TO 44	1 TO 10			
1905	129 TO 247	87 TO 163	45 TO 95	11 TO 26			
1906	248 TO 385	164 TO 223	96 TO 162				
1907[a]	386 TO ?	224 TO 302	163 TO 215			387 TO 396	241 TO 250
1908[a]	???	303 TO 377	216 TO 262		1 TO 12 (HEAVY) 13 TO 24		
1909[a]	? TO 746	378 TO 477	263 TO 323				
1910	747 TO 928	478 TO 552	324 TO 379	1 TO 10			

[1] Sleighs were built by the Grubers, but the information related to this product is confusing and some serial numbers appear to have been duplicated.

[a] Wagon record numbers are intermingled throughout these years. See actual records for accurate listings.

Table IV. Production Records for the Years 1911–1934.

1911

RUNNING GEARS

SERIAL NUMBER	AXLE SIZE	SERIAL NUMBER	AXLE SIZE
929	2 1/8"	948	2"
930	2 1/8"	949	2"
931	1 7/8"	950	2"
932	1 3/4"	951	2"
933	2 1/4"	952	2"
934	2 1/8"	953	2"
935	2 1/8"	954	2 1/8"
936	1 3/4"	955	2 1/8"
937	1 3/4"	956	
938	1 3/4"	957	1 3/4"
939	1 3/4"	958	1 3/4"
940	1 3/4"	959	2 1/4"
941	1 3/4"	960	1 5/8"
942	1 3/4"	961	1 5/8"
943	1 3/4"	962	1 3/8"
944	1 3/4"	963	1 3/8"
945	1 3/4"	964	1 1/2"
946	1 3/4"	965	2 1/8"
947	1 3/4"	966	2"

Table IV, continued; continues 1911 Running Gears.

SERIAL NUMBER	AXLE SIZE	SERIAL NUMBER	AXLE SIZE
967	2"	1005	1 1/2" AXLE 1 3/8" GEAR
968	2"	1006	1 1/2" AXLE 1 3/8" GEAR
969	2"	1007	1 3/8"
970	2"	1008	1 3/4"
971	2"	1009	1 3/4"
972	2"	1010	2 1/4" AXLE 2" GEAR
973	1 7/8"	1011	1 3/4"
974	1 7/8"	1012	1 3/4"
975	1 7/8"	1013	1 3/4"
976	1 7/8"	1014	1 3/4"
977	1 3/4"	1015	1 3/4"
978	1 5/8"	1016	1 3/4"
979	2 1/4"	1017	1 3/4"
980	1 3/4"	1018	1 3/4" ENGINE WAGON
981	1 3/4"	1019	2"
982	1 3/4"	1020	2"
983	1 3/4"	1021	2"
984	1 3/4"	1022	2"
985	1 3/4"	1023	2"
986	1 3/4"	1024	2"
987	1 5/8"	1025	2"
988	1 5/8"	1026	2"
989	1 5/8"	1027	2"
990	1 5/8"	1028	2"
991	1 5/8"	1029	2"
992	1 5/8"	1030	2"
993	1 5/8"	1031	1 3/4" GEAR 1 7/8" AXLE
994	1 5/8"	1032	1 1/2" GEAR 1 5/8" AXLE
995	1 5/8"	1033	1 1/4"
996	1 7/8" CART	1034	1 1/4"
997	2 1/8"	1035	1 1/4"
998	2 1/8"	1036	1 1/4"
999	2 1/8"	1037	1 1/4"
1000	2 1/8"	1038	1 1/4"
1001	2 1/8"	1039	1 3/8"
1002	2 1/8"	1040	1 3/8"
1003	1 7/8" NEIN BROS	1041	1 3/8"
1004	2 1/8" KIRSHMAN	1042	1 3/8"

BOXES

SERIAL NUMBER	AXLE SIZE	SERIAL NUMBER	AXLE SIZE
553	2 1/8"	560	1 5/8"
554	2 1/8"	561	1 5/8"
555	2"	562	1 5/8"
556	1 1/2"	563	1 5/8"
557	1 3/4"	564	1 5/8"
558	1 5/8"	565	1 5/8"
559	1 5/8"	566	1 5/8"

Table IV, continued; continues 1911 Boxes.

SERIAL NUMBER	AXLE SIZE	SERIAL NUMBER	AXLE SIZE
567	1 5/8"	609	2"
568	1 5/8"	610	2"
569	1 5/8"	611	2"
570	1 5/8"	612	1 5/8"
571	1 5/8"	613	1 3/4" NEIN BROS
572	1 5/8"	614	1 3/4" - 3'-6"
573	1 5/8"	615	1 3/8" AXLE 1 1/2" SHAFTS
574	1 1/2"	616	1 3/8" AXLE 1 1/2" SHAFTS
575	1 1/2"	617	1 7/8"
576	1 1/2"	618	1 3/4"
577	1 1/2"	619	1 3/4"
578	1 1/2"	620	1 3/4"
579	1 3/4"	621	1 3/4"
580	1 3/4"	622	1 3/4"
581	1 7/8"	623	1 3/4"
582	1 3/4"	624	1 3/4"
583	1 3/4"	625	1 3/4"
584	1 3/4"	626	1 3/4"
585	1 3/4"	627	1 3/4"
586	1 5/8"	628	1 3/4"
587	1 5/8"	629	1 3/4"
588	1 5/8"	630	1 3/4"
589	1 5/8"	631	1 3/4"
590	1 5/8"	632	1 5/8"
591	2"	633	1 5/8"
592	2"	634	1 5/8"
593	2"	635	1 5/8"
594	2"	636	1 5/8"
595	2"	637	1 5/8"
596	1 3/4"	638	1 5/8"
597	1 3/4"	639	1 5/8"
598	1 3/4"	640	1 5/8"
599	1 3/4"	641	1 5/8"
600	1 3/4"	642	1 5/8"
601	1 3/4"	643	1 5/8"
602	1 7/8"	644	1 5/8"
603	1 3/8"	645	1 5/8"
604	1 3/8"	646	1 5/8"
605	1 1/2"	647	1 5/8"
606	1 3/4"	648	1 1/2" SHAFTS
607	1 3/4"	649	1 1/2" ROTH'S
608		650	1 7/8"

HAY FLATS

380	2"	384	2"
381	2"	385	1 5/8"
382	1 3/4"	386	2 1/8"
383	2 1/4"	387	1 3/4"

Table IV, continued; continues 1911 Hay Flats.

SERIAL NUMBER	AXLE SIZE	SERIAL NUMBER	AXLE SIZE
388	2"	408	2"
389	2"	409	2"
390	2"	410	1 3/4"
391	2"	411	1 3/4"
392	2"	412	2"
393	2"	413	1 7/8"
394	2"	414	2"
395	2"	415	2"
396	1 3/4"	416	2 1/4"
397	2 1/8"	417	2"
398	2 1/4"	418	2"
399	1 5/8"	419	2"
400	1 5/8"	420	2"
401	2 1/4"	421	2"
402	1 5/8"	422	2"
403	1 1/2"	423	2"
404	1 3/4"	424	2" -18"
405	2"	425	2" -18"
406	2"	426	1 5/8" & 1 3/4" - 16'
407	1 5/8"	427	1 3/4" - 16'

1912

RUNNING GEARS

1042	1 3/8"	1065	1 1/2"
1043	2 1/4"	1066	1 1/2"
1044	2 1/8"	1067	1 1/2" ROTHS
1045	2 1/8"	1068	1 5/8" ROTHS
1046	2 1/8"	1069	1 5/8"
1047	1 5/8"	1070	1 3/4"
1048	1 5/8"	1071	2" AXLE 1 7/8" GEAR
1049	1 5/8"	1072	2" AXLE 1 7/8" GEAR
1050	1 5/8"	1073	1 5/8"
1051	1 5/8"	1074	1 3/4" DUNDORE
1052	1 5/8"	1075	2 1/8" SHEIDY
1053	1 5/8"	1076	
1054	1 5/8"	1077	1 7/8"
1055	1 3/4"	1078	1 7/8"
1056	1 1/8" AXLE 1 3/4" GEAR	1079	1 7/8"
1057	2"	1080	2"
1058	2"	1081	2"
1059	2"	1082	2"
1060	1 1/2"	1083	1 3/4"
1061	1 1/2"	1084	1 3/4"
1062	1 1/2"	1085	1 1/2" AXLE 1 3/8" GEAR
1063	1 1/2"	1086	1 3/4"
1064	1 1/2"	1087	1 3/4"

Table IV, continued; continues 1912 Running Gears.

SERIAL NUMBER	AXLE SIZE	SERIAL NUMBER	AXLE SIZE
1088	2" AXLE 1 7/8" GEAR	1112	1 3/4"
1089	1 5/8"	1113	1 3/4"
1090	1 5/8"	1114	1 3/4"
1091	1 5/8"	1115	1 3/4"
1092	1 5/8"	1116	1 3/4"
1093	1 5/8"	1117	1 3/4"
1094	1 5/8"	1118	1 3/4"
1095	1 5/8"	1119	1 3/4"
1096	1 5/8"	1120	1 3/4"
1097	1 1/2" AXLE 1 3/8" GEAR	1121	1 3/4"
1098	1 1/2" AXLE 1 3/8" GEAR	1122	1 3/4"
1099	5/8" SMALL WAGONS	1123	1 3/4"
1100	5/8" SMALL WAGONS	1124	1 3/4"
1101	1 5/8"	1125	1 3/4"
1102	1 5/8"	1126	1 3/4"
1103	1 5/8"	1127	1 3/4"
1104	1 5/8"	1128	1 3/4"
1105	1 5/8"	1129	1 3/4"
1106	1 5/8"	1130	1 3/4"
1107	1 5/8"	1131	1 3/4"
1108	1 5/8"	1132	1 3/4"
1109	1 5/8"	1133	1 7/8"
1110	1 3/4"	1134	1 1/2" ROTHS
1111	1 3/4"	1135	1 1/2" AXLE 1 3/8" GEAR

BOXES

SERIAL NUMBER	AXLE SIZE	SERIAL NUMBER	AXLE SIZE
650	1 7/8"	672	1 5/8"
651	1 1/4"	673	1 5/8"
652	1 1/4"	674	1 1/2"
653	1 1/4"	675	1 1/2"
654	1 1/4"	676	1 1/2"
655	1 1/4"	678	1 1/2"
656	1 1/4"	679	1 1/2"
657	1 3/8"	680	1 1/2"
658	1 3/8"	681	1 1/2" ROTH'S
659	1 3/8"	682	1 5/8" 10' LONG
660	1 3/8"	683	2" STONE BOX
661	1 3/4" 18"-8"	684	2"
662	1 3/4" 16"-8"	685	2"
663	2"	686	2"
664	2"	687	2"
665	2"	688	1 7/8"
666	1 7/8" FISHER	689	1 7/8"
667	1 5/8" MILL BOX	690	1 3/4"
668	1 5/8"	691	1 1/2" SHAFTS &TONGUE
669	5/8"	692	1 3/4" WEBBER
670	1 5/8"	693	1 3/4" WEBBER
671	1 5/8"	694	1 1/2" BRICKER

Table IV, continued; continues 1912 Boxes.

SERIAL NUMBER	AXLE SIZE	SERIAL NUMBER	AXLE SIZE
695	1 1/2" HART	703	1 5/8"
696	SMALL WAGON	704	1 5/8"
697	SMALL WAGON	705	1 5/8"
698	1 3/4" NEIN BROS	706	1 5/8"
699	1 3/4"	707	1 5/8"
700	1 3/4" 10'-6" FLOOR	708	1 1/2" ROTH'S
701	1 3/4"	709	
702	1 5/8"		

HAY FLATS

427	1 3/4" - 16'	454	2" - 18'
428	2" - 16'	455	2" - 18'
429	2" - 18'	456	1 7/8" - 18'
430	1 3/4" - 14'	457	1 5/8" - 16'
431	2" - 18'	458	1 1/2" - 14'
432	1 7/8" - 18'	459	2" - 18'
433	2 1/8" - 18'	460	2 1/8" - 18'
434	2 1/8" - 18'	461	2" - 16'
435	1 5/8" - 14' STEP, SEAT, BRAKE	462	2 1/4" - 18'
436	2 1/8" - 18'	463	2" - 16'
437	1 3/4" - 16'	464	2" - 18'
438	2" - 18'	465	2" - 20'
439	2" - 18'	466	2 1/8" - 18'
440	2" - 18'	467	2" - 16'
441	2" - 16'	468	2" - 18'
442	2" - 16'	469	2" - 18'
443	2 1/8" - 18'	470	1 5/8" - 14'
444	1 7/8" - 18'	471	1 5/8" - 14'
445	1 3/4" - 14'	472	2" - 18'
446	2" - 16'	473	2 1/4" - 18'
447	2" - 18'	474	2 1/8" - 18'
448	2" - 18'	475	2 1/4" - 18'
449	1 5/8" - 16'	476	2 3/4" - 20'
450	1 5/8" - 16'	477	1 3/4" - 16'
451	1 5/8" - 14'	478	1 3/4" - 14'
452	2 1/8" - 20'	479	2" - 16'
453	2" - 18'		

1913

RUNNING GEARS

1136	1143
1137	1144
1138	1145
1139	1146
1140	1147
1141	1148
1142	1149

Table IV, continued; continues 1913 Running Gears.

SERIAL NUMBER	AXLE SIZE	SERIAL NUMBER	AXLE SIZE
1150		1198	2 1/4"
1151		1199	1 3/4"
1152		1200	1 5/8"
1153		1201	1 5/8"
1154		1203	1 1/2"
1155		1204	2 1/4" - 2" GEAR
1156		1205	1 1/2" - 1 3/8" GEAR
1157		1206	1 1/2" - 1 3/8" GEAR
1158		1207	1 1/2"
1159		1208	1 3/4"
1160		1209	1 3/4"
1161		1210	1 3/8"
1162		1211	1 3/8"
1163		1212	2"
1164		1213	2"
1165		1214	2" CORNER CHURCH
1166		1215	1 3/4"
1167		1216	1 3/4"
1168		1217	1 7/8"
1169		1218	1 5/8"
1170		1219	2"
1171		1220	2"
1172		1221	2"
1173		1222	2"
1174		1223	2 1/8" ROTH'S
1175		1224	1 3/4"
1176		1225	1 3/4"
1177		1226	1 3/4"
1178		1227	1 7/8" - 1 3/4" GEAR
1179		1228	2 1/8"
1180		1229	2 1/8"
1181		1230	2 1/8"
1182		1231	2"
1183		1232	2"
1184	1 3/4" - 1 5/8" GEAR	1233	2"
1185	1 5/8"	1234	2"
1186	1 5/8"	1235	1 5/8"
1187	1 3/4" - 1 5/8" GEAR	1236	1 5/8"
1188	1 3/4"	1237	1 5/8"
1189	2"	1238	1 5/8"
1190	2"	1239	1 5/8"
1191	2"	1240	1 5/8"
1192	2"	1241	1 5/8"
1193	2"	1242	1 5/8"
1194	2"	1243	1 5/8"
1194	2"	1244	1 5/8"
1196	2"	1245	1 5/8"
1197	2 1/8"	1246	1 5/8"

Table IV, continued; continues 1913 Running Gears.

SERIAL NUMBER	AXLE SIZE	SERIAL NUMBER	AXLE SIZE
1247	1 5/8"	1261	1 1/2" - 1 3/8" GEAR
1248	1 5/8"	1262	1 3/4"
1249	1 1/4"	1263	1 3/4"
1250	1 1/4"	1264	1 3/4"
1251	1 1/4"	1265	1 3/4"
1252	1 5/8"	1266	1 3/4"
1253	1 5/8"	1267	1 3/4"
1254	1 5/8"	1268	1 3/4"
1255	1 5/8"	1269	1 3/4"
1256	1 1/2"	1270	1 3/4"
1257	1 1/2"	1271	1 3/4"
1258	1 1/2"	1272	1 3/4"
1259	1 1/2"	1273	1 3/4"
1260	1 1/2" - 1 3/8" GEAR	1274	1 3/4"

BOXES

710	1 1/2" SHAFTS	742	1 1/2"
711	1 3/4"	743	1 3/4"
712	1 3/4"	744	1 3/4"
713	1 3/4"	745	1 3/4"
714	1 3/4"	746	1 3/4"
715	1 3/4"	747	1 5/8"
716	1 3/4" STONE BED	748	1 5/8"
717	1 1/4"	749	1 5/8"
718	1 1/4"	750	
719	1 1/4"	751	
720	1 3/8"	752	1 3/4"
721	1 3/8"	753	1 3/4"
723	1 3/8"	754	2"
724	1 5/8"	755	2"
725	1 5/8"	756	2"
726	1 5/8"	757	1 1/2" SHAFTS
727	1 5/8"	758	
728	1 5/8"	759	2 1/4"
729	1 5/8"	760	1 3/4"
730	1 5/8"	761	1 3/4"
731	1 5/8"	762	1 3/4"
732	1 5/8"	763	1 3/4"
733	1 5/8"	764	CORNER CHURCH
734	1 3/4"	765	1 3/8"
735	1 3/4"	766	1 3/8"
736	1 3/4"	767	1 3/4" OLD WAGON
737	1 3/4"	768	1 3/4"
738	1 3/4"	769	1 3/8"
739	1 3/4"	770	1 7/8" STONE BED
740	1 1/2" ROTH'S	771	2 1/8" ROTH'S
741	1 3/4"	772	1 5/8"

Table IV, continued; continues 1913 Boxes.

SERIAL NUMBER	AXLE SIZE	SERIAL NUMBER	AXLE SIZE
773	1 5/8"	776	1 5/8"
774	1 5/8"	777	1 5/8"
775	1 5/8"	778	

HAY FLATS

SERIAL NUMBER	AXLE SIZE	SERIAL NUMBER	AXLE SIZE
480	2" - 18'	511	2" - 18'
481	2" - 18'	512	2 1/8" - 18'
482	2" - 18'	513	2"
483	2 1/4" - 18'	514	2"
484	2" - 18'	515	2"
485	2" - 18'	516	1 5/8"
486	1 5/8" - 14'	517	2"
487	1 3/4" - 14'	518	2 1/4" - 18'
488	1 3/4" - 14'	519	2" - 16'
489	2" - 18'	520	2" - 16'
490	2" - 18'	521	2" - 18'
491	2" - 18'	522	2" - 16'
492	1 3/4" - 16'	523	2 1/4" - 18'
493	1 5/8" - 16'	524	1 3/4" - 18'
494	1 3/4" - 14'	525	2" - 18'
495	1 3/4" - 16'	526	2" - 18'
496	2 1/8" - 18'	527	2" - 18'
497	2 1/8" - 18'	528	2"
498	1 7/8" - 18'	529	2" - 16'
499		530	2" - 16'
500		531	2" - 18'
501	2" - 16'	532	2" - 16'
502	2" - 16'	533	2 1/4" - 18'
503	2" - 20'	534	2" - 18'
504	2" - 18'	535	1 3/4" - 16'
505	2" - 16'	536	1 7/8" - 16'
506	2" - 18'	537	1 3/4" - 14'
507	2" - 16'	538	1 3/4" - 12'
508	2 1/8" - 18'	539	1 3/4" - 12'
509	2" - 18'	540	2" - 16'
510	2" - 18'	541	2" - 18'

1914

RUNNING GEARS

SERIAL NUMBER	AXLE SIZE	SERIAL NUMBER	AXLE SIZE
1275	1 7/8"	1284	2 1/8"
1276	1 7/8"	1285	2"
1277	1 7/8"	1286	2"
1278	1 7/8"	1287	2"
1279	1 7/8"	1288	2"
1280	1 7/8"	1289	2"
1281	1 7/8"	1290	2"
1282	1 3/8"	1291	2"
1283	1 7/8" - 1 3/4" GEAR	1292	2"

Table IV, continued; continues 1914 Running Gears.

SERIAL NUMBER	AXLE SIZE	SERIAL NUMBER	AXLE SIZE
1293	2"	1341	1 3/4"
1294	2"	1342	1 3/4"
1295	2"	1343	1 3/4"
1296	2 1/2"	1344	1 3/8"
1297	2"	1345	1 3/8"
1298	1 3/4"	1346	1 3/8"
1299	1 3/4"	1347	1 1/4"
1300	1 3/4"	1348	1 1/4"
1301	1 3/4"	1349	1 1/4"
1302	1 3/4"	1350	2"
1303	1 3/4"	1351	2"
1304	1 3/4"	1352	2"
1305	1 1/2"	1353	2"
1306	1 1/2"	1354	2"
1307	2 1/4"	1355	2"
1308	2 1/8"	1356	2"
1309	2 1/8"	1357	1 1/2" ED BOHN
1310	2 1/8" GRING	1358	1 5/8"
1311	1 5/8"	1359	1 5/8"
1312	1 5/8"	1360	1 5/8"
1313	2 1/8"	1361	1 5/8"
1314	1 5/8"	1362	1 5/8"
1315	1 5/8"	1363	1 5/8"
1316	1 5/8"	1364	1 5/8"
1317	1 5/8"	1365	1 5/8"
1318	1 5/8"	1366	1 5/8"
1319	1 5/8"	1367	1 5/8"
1320	1 5/8"	1368	1 5/8"
1321	2 1/8"	1369	1 5/8"
1322	2"	1370	1 5/8"
1323	2"	1371	1 5/8"
1324	2"	1372	1 5/8"
1325	2"	1373	1 5/8"
1326	2" DEGLER	1374	1 5/8"
1327	2 1/8" D. KLEIN	1375	1 5/8"
1328	1 3/4"	1376	1 5/8"
1329	1 5/8"	1377	1 5/8"
1330	1 5/8"	1378	2 1/8"
1331	1 5/8"	1379	1 3/4"
1332	1 5/8"	1380	1 3/4"
1333	1 5/8"	1381	2" ROTH'S
1334	1 5/8"	1382	1 5/8" ROTH'S
1335	1 7/8"	1383	2 1/4"
1336	1 3/4"	1384	1 3/8"
1337	1 3/4"	1385	2 3/4" KALBACH
1338	1 3/4"	1386	2 1/2"
1339	1 3/4"	1387	1 1/4"
1340	1 3/4"	1388	1 1/2"

Table IV, continued; continues 1914 Running Gears.

SERIAL NUMBER	AXLE SIZE	SERIAL NUMBER	AXLE SIZE
1389	1 1/2"	1392	1 1/2"
1390	1 1/2"	1393	1 1/2"
1391	1 1/2"	1394	1 1/2"

BOXES

SERIAL NUMBER	AXLE SIZE	SERIAL NUMBER	AXLE SIZE
779	1 5/8"	820	10' - 6"
780	1 5/8"	821	10' - 6"
781	1 5/8"	822	1 1/2"
782	1 5/8"	823	1 1/2"
783	1 5/8"	824	1 5/8"
784	1 5/8"	825	2" ROTH'S
785	1 5/8"	826	
786	1 5/8"	827	
787	1 5/8"	828	1 1/2"
788	1 5/8"	829	1 3/4"
789	1 3/4"	830	1 3/4"
790	1 3/4"	831	1 5/8"
791	1 3/4"	832	1 5/8"
792	1 3/4"	833	1 5/8"
793	1 3/4"	834	1 5/8"
794	1 3/4"	835	1 5/8"
795	1 3/4"	836	1 5/8"
796	1 3/4"	837	
797	1 3/4"	838	
798	1 3/4"	839	
799	1 1/2"	840	
800	1 1/2"	841	
801	1 1/2"	842	
802	1 1/2"	843	2"
803	1 1/2"	844	2"
804	SHELVES	845	
805	1 3/8"	846	
806		847	
807	1 1/4"	848	
808	1 1/4"	849	
809	1 1/4"	850	
810	1 3/4"	851	
811	1 3/8"	852	
812	2"	853	
813	2"	854	2" ROTH'S
814		855	1 5/8" ROTH'S
815		856	1 3/8"
816		857	
817		858	
818	1 3/4"	859	1 3/8"
819	10' - 6"		

Table IV, continued; begins 1914 Hay Flats.

SERIAL NUMBER	AXLE SIZE	SERIAL NUMBER	AXLE SIZE
HAY FLATS			
542	1 3/4" - 14'	571	2" - 18'
543	2" - 18'	572	2" - 18'
544	2" - 18'	573	2 1/8" - 18'
545	1 3/4" - 14'	574	2" - 18'
546	2 1/8" - 18'	575	
547	2" - 18'	576	
548	2" - 18'	577	1 3/4" - 16'
549	2 1/8" - 18'	578	1 5/8"
550	2" - 18'	579	2" - 18'
551	2" - 18'	580	2 1/4" - 18'
552	2" - 18'	581	2" - 16'
553	1 3/4" - 14'	582	2 1/8" - 18'
554	2 1/8" - 18'	583	2 1/8" - 18'
555	1 3/4" - 16'	584	1 7/8" - 18'
556	1 5/8" - 14'	585	2 1/8" - 18'
557	1 7/8" - 18'	586	1 5/8" - 14'
558	2" - 18'	587	2" - 18'
559	2 1/8" - 18'	588	2" - 16'
560	2" - 18'	589	2" - 18'
561	1 1/2" - 14'	590	1 3/4" - 14'
562	1 5/8" - 14'	591	2 1/8" - 18'
563	2" - 18'	592	2 1/8" - 18'
564	2" - 18'	593	2 1/4" - 18'
565	2 1/4" - 18'	594	2" - 18'
566	2 1/4" - 18'	595	2" - 18'
567	1 3/4" - 16'	596	2" - 18'
568	2" - 18'	597	
569	1 3/4" - 14'	598	
570	2" - 18'	599	

1915

RUNNING GEARS

SERIAL NUMBER	AXLE SIZE	SERIAL NUMBER	AXLE SIZE
1395	2"	1409	2 1/4"
1396	2"	1410	1 3/4"
1397	2"	1411	1 3/4"
1398	2"	1412	1 3/4"
1399	2 1/8"	1413	1 3/4"
1400	2 1/8"	1414	1 3/4"
1401	2 1/8"	1415	1 3/4"
1402	2 1/8"	1416	1 3/4"
1403	1 3/8"	1417	1 3/4"
1404	1 3/8"	1418	1 3/4"
1405	1 3/8"	1419	1 3/4"
1406	1 5/8" C. D. MILLER	1420	1 3/4"
1407	1 3/4"	1421	1 3/4" - 1 7/8" GEAR
1408	2"	1422	1 3/4"

Table IV, continued; continues 1915 Running Gears.

SERIAL NUMBER	AXLE SIZE	SERIAL NUMBER	AXLE SIZE
1423	1 3/4"	1460	1 3/4"
1424	1 7/8"	1461	MERRIT
1425	1 7/8"	1462	LONG
1426	2 1/8"	1463	1 3/8"
1427	2 1/8"	1464	1 5/8" HAMBURG
1428	2"	1465	1 7/8" - 1 3/4"GEAR
1429	2 1/8"	1466	1 3/8"
1430	1 1/2" ROTH'S	1467	1 3/8"
1431	2 1/8" VALENTINE	1468	1 3/8"
1432	1 7/8"	1469	1 3/8"
1433	1 7/8"	1470	1 3/8"
1434	1 7/8"	1471	1 3/8"
1435	2"	1472	3 1/2" TRAILER
1436	2"	1473	1 7/8" SNYDER
1437	2"	1474	1 5/8"
1438	2"	1475	1 5/8"
1439	2"	1476	1 5/8"
1440	2"	1477	1 5/8"
1441	2"	1478	1 5/8"
1442	2"	1479	1 5/8"
1443	2"	1480	1 5/8"
1444	2" STONE WAGON	1481	1 5/8"
1445	2 1/8"	1482	1 5/8"
1446	2 1/8"	1483	1 5/8"
1447	2 1/8"	1484	1 1/2" ROTH'S
1448	2 1/8"	1485	2 1/4"
1449	1 3/4"	1486	1 3/4"
1450	1 3/4"	1487	1 3/4"
1451	1 3/4"	1488	1 3/4"
1452	1 3/4"	1489	1 3/4"
1453	1 3/4"	1490	1 3/4"
1454	1 3/4"	1491	1 3/4"
1455	1 3/4"	1492	1 3/4"
1456	1 3/4"	1493	1 3/4"
1457	1 3/4"	1493	1 3/4"
1458	1 3/4"	1494	1 3/4"
1459	1 3/4"	1495	1 3/4"

BOXES

SERIAL NUMBER	AXLE SIZE	SERIAL NUMBER	AXLE SIZE
860	1 5/8"	870	1 5/8"
861	1 5/8"	871	1 5/8"
862	1 5/8"	872	1 3/4"
863	1 5/8"	873	1 3/4"
864	1 5/8"	874	1 3/4"
865	1 5/8"	875	1 3/4"
866	1 5/8"	876	1 3/4"
867	1 5/8"	877	1 3/4"
868	1 5/8"	878	1 3/4"
869	1 5/8"	879	1 3/4"

Table IV, continued; continues 1915 Boxes.

SERIAL NUMBER	AXLE SIZE	SERIAL NUMBER	AXLE SIZE
880	1 3/4"	909	1 5/8"
881	1 3/4"	910	1 7/8"
882	2"	911	1 7/8"
883	2"	912	2"
884	2"	913	2"
885	1 1/2"	914	HAMBURG
886	1 1/2"	915	1 3/8"
887	1 1/2"	916	1 3/8"
888	1 1/2"	917	1 3/8"
889	1 1/2" ROTH'S	918	1 3/8"
890	1 3/8" BENDER	919	1 3/8"
891	1 1/2" ROTH'S	920	1 3/8"
892	1 3/4" COAL BOX	921	1 7/8" SNYDER
893	2 1/4" STONE BED	922	1 5/8" ROTH'S
894	2 1/8" STONE BED	923	
895	1 7/8" COAL BOX (LONG)	924	1 3/4"
896	1 3/8"	925	1 3/4"
897	1 3/8"	926	1 3/4"
898	1 3/4"	927	1 3/4"
899	1 3/4"	928	1 3/4"
900	1 3/4"	929	1 3/4"
901	1 3/4"	930	1 3/4"
902	1 3/4"	931	1 3/4"
903	1 3/4"	932	1 3/4"
904	1 5/8"	933	1 3/4"
905	1 5/8"	934	1 3/4"
906	1 5/8"	935	1 3/4"
907	1 5/8"	936	1 3/4"
908	1 5/8"	937	1 3/4"

HAY FLATS

600	2" - 18'	617	2 1/8" - 18'
601	2" - 16'	618	2 1/8" - 18'
602	1 3/4" - 14'	619	2" - 18' CUMMINGS
603	2 1/8" - 18'	620	2 1/8" - 18'
604	2 1/4" - 20'	621	1 7/8" - 16' STEP, SEAT, BRAKE
605	2" - 18'	622	1 7/8" - 16' STEP, SEAT, BRAKE
606	2" - 18'	623	1 5/8" - 14'
607	2 1/8" - 18'	624	2" - 18'
608	2" - 18'	625	2" - 18'
609	1 3/4" - 14'	626	2" - 18'
610	1 3/4" - 14'	627	18' ACME PAPER CO
611	2" - 16'	628	2 1/8" - 18'
612	2" - 18' STEP,SEAT,BRAKE	629	1 3/4" - 14'
613	2" - 18'	630	1 5/8" - 16'
614	1 3/4" - 16'	631	2" - 20'
615	1 7/8" - 18'	632	2 1/8" - 18'
616	1 7/8" - 16'	633	

Table IV, continued; continues 1915 Hay Flats.

SERIAL NUMBER	AXLE SIZE	SERIAL NUMBER	AXLE SIZE
634	DIDN'T EXIST	652	1 3/4" - 14'
635	2" - 18'	653	1 3/4" - 16'
636	2" - 18'	654	1 3/4" - 14'
637	1 7/8" - 16'	655	2 1/8" - 18'
638	2 1/8" - 18'	656	2" - 18'
639	2" - 18'	657	1 5/8" - 16'
640	2" - 18'	658	2" - 18'
641	2" - 18'	659	1 3/4" - 16'
642	2" - 16'	660	1 3/4" - 16'
643	1 3/4" - 16'	661	1 3/4" - 14'
644	2" - 16'	662	1 3/4" - 14'
645	2" - 16'	663	1 3/4" - 12'
646	2 1/4" - 18'	664	2 1/8" - 18'
647	2"	665	2" - 18'
648	2"	666	2" - 18'
649	2"	667	2 1/4" - 18'
650		668	1 3/4" - 14'
651			

1916

RUNNING GEARS

1496	2"	1523	1 3/4"
1497	2"	1524	1 3/4"
1498	2"	1525	1 3/4"
1499	2"	1526	1 3/4"
1500	2"	1527	1 3/4"
1501	2"	1528	1 3/4"
1502	2"	1529	1 3/4"
1503	2"	1530	2"
1504	2"	1531	2"
1505	2"	1532	2"
1506	2"	1533	2"
1507	2"	1534	2"
1508	2"	1535	2"
1509	2"	1536	2"
1510	2"	1537	2"
1511	2 1/8"	1538	2"
1512	2 1/8"	1539	2"
1513	2 1/8"	1540	2 1/8"
1514	1 1/2" J. M. GRING	1541	2 1/8"
1515	1 7/8" - 1 3/4"GEAR (HAIN)	1542	2 1/8"
1516	2 1/4"	1543	2 1/8"
1517	2 1/4"	1544	2 1/8"
1518	1 1/4"	1545	1 3/4"
1519	1 1/4"	1546	1 3/4"
1520	1 1/4"	1547	2 1/8"
1521	1 3/4"	1548	2 1/8"
1522	1 3/4"	1549	1 3/4" PEIFFER

Table IV, continued; continues 1916 Running Gears.

SERIAL NUMBER	AXLE SIZE	SERIAL NUMBER	AXLE SIZE
1550	1 7/8" RUTH	1572	2 1/8" ROTH'S
1551	2"	1573	2 1/8" ROTH'S
1552	2"	1574	1 5/8"
1553	2"	1575	1 5/8"
1554	2"	1576	1 5/8"
1555	2 1/8"	1577	1 5/8"
1556	2 1/8"	1578	1 5/8"
1557	1 5/8"	1579	1 5/8"
1558	1 5/8"	1580	1 5/8"
1559	1 5/8"	1581	1 5/8"
1560	1 5/8"	1582	1 5/8"
1561	1 5/8"	1583	1 5/8"
1562	1 5/8"	1584	1 5/8"
1563	1 5/8"	1585	1 5/8"
1564	1 5/8"	1586	1 3/4"
1565	1 5/8"	1587	1 3/4"
1566	1 5/8"	1588	1 3/4"
1567	1 5/8" DUBBS & ZUNN	1589	1 3/4"
1568	1 5/8" GARMAN BROS	1590	1 3/4"
1569	1 7/8" GASS BROS	1591	1 3/4"
1570	1 5/8" ROTH'S (I. MILLER)	1592	1 3/4"
1571	1 5/8" ROTH'S (I. MILLER)	1593	1 3/4"

BOXES

SERIAL	AXLE SIZE	SERIAL	AXLE SIZE
938	1 5/8"	960	1 1/2"
939	1 5/8"	961	1 1/2"
940	1 5/8"	962	1 1/2"
941	1 5/8"	963	1 1/2"
942	1 5/8"	964	1 5/8" LINCOLN
943	1 5/8"	965	1 7/8" GASS BROS
944	1 5/8"	966	1 5/8" ROYER
945	1 5/8"	967	1 5/8" DUBBS & ZUNN
946	1 5/8"	968	2 1/8" MILLER
947	1 5/8"	969	2 1/8" MILLER
948	1 5/8"	970	1 5/8" ROTH'S (I. MILLER)
949	1 5/8"	971	1 5/8" ROTH'S (I. MILLER)
950	1 5/8"	972	2 1/8" FISHER
951	1 5/8"	973	TRUCK BODY
952	1 5/8" J.M. GRING	974	1 3/4" WAGNER
953	1 5/8" OLD WAGON	975	1 3/4"
954	1 3/4" OLD WAGON	976	1 3/4"
955	1 3/8"	977	1 3/4"
956	1 1/4"	978	1 3/4"
957	1 1/4"	979	1 3/4"
958	1 1/4"	980	1 3/4"
959	1 3/4" GASS OLD WAGON		

Table IV, continued; begins 1916 Hay Flats.

SERIAL NUMBER	AXLE SIZE	SERIAL NUMBER	AXLE SIZE
	HAY FLATS		
669	2 1/4" - 16'	698	2 1/8" - 18'
670	2" - 18' DUBBS	699	2" - 18'
671	2" - 18'	700	2 1/8" - 18'
672	2" - 18'	701	2" - 18'
673	1 3/4" - 16' GULDIN	702	2" - 18'
674	2 1/8" - 16'	703	2" - 18'
675	1 7/8" - 16' O. RUTH	704	1 3/4" - 16'
676	1 7/8" - 14' HIMMELBERGER	705	1 3/4" - 16'
677	2 1/8" - 18'	706	1 3/4" - 16'
678	2" - 18' STOLTZFUS	707	2 1/8" - 18'
679	2 1/8" - 18'	708	1 5/8" - 16'
680	2" - 18'	709	1 3/4" - 16'
681	2" - 18'	710	1 3/4" - 14'
682	2 1/8" - 18'	711	2" - 18'
683	2" - 16'	712	2" - 16'
684	1 7/8" - 18' RUTH	713	1 5/8" - 16'
685	2" - 16'	714	2 1/8" - 18'
686	1 3/4" - 16'	715	1 7/8" - 18'
687	2" - 18'	716	2 1/4" - 18'
688	2 1/8" - 16'	717	1 3/4" - 16'
689	2 1/4" - 18' ROYER	718	2" - 18'
690	2" - 18'	719	1 5/8" - 14'
691	1 3/4" - 16'	720	2" - 18'
692	1 3/4" - 16'	721	2" - 18'
693	1 3/4" - 18'	722	2 1/8" - 18'
694	1 3/4" - 16' OLD WAGON	723	2" - 18'
695	2" - 18'	724	2" - 16'
696	2" - 16'	725	1 3/4" - 16'
697	2 1/8" - 18'		

1917

RUNNING GEARS

1594	1 1/2"	1608	2"
1595	1 1/2"	1609	2"
1596	1 1/2"	1610	2"
1597	1 1/2"	1611	1 3/4"
1598	1 1/2"	1612	1 3/8"
1599	1 1/2"	1613	1 7/8"
1600	2 1/4"	1614	1 7/8"
1601	1 7/8"	1615	1 3/4" DEGLER
1602	1 7/8"	1616	1 3/4" WEITZEL
1603	1 7/8"	1617	1 3/4"
1604	1 7/8"	1618	1 3/4"
1605	2"	1619	1 3/4"
1606	2"	1620	1 3/4"
1607	2"	1621	1 7/8"

Table IV, continued; continues 1917 Running Gears.

SERIAL NUMBER	AXLE SIZE	SERIAL NUMBER	AXLE SIZE
1622	1 3/4"	1649	1 3/4"
1623	2 1/4"	1650	1 3/4"
1624	2 1/8"	1651	1 5/8"
1625	2"	1652	1 3/4"
1626	2"	1653	1 3/4"
1627	2"	1654	1 3/4"
1628	2"	1655	1 1/2" - 1 3/8" GEAR
1629	2"	1656	1 7/8" - 1 3/4" GEAR
1630	2"	1657	1 5/8"
1631	1 3/4"	1658	1 5/8"
1632	1 3/4"	1659	1 5/8"
1633	1 3/4"	1660	1 3/4"
1634	1 3/4"	1661	1 5/8"
1635	1 7/8" NEIN	1662	1 5/8"
1636	1 3/8"	1663	1 5/8"
1637	1 3/8"	1664	1 5/8"
1638	1 3/8"	1665	1 5/8"
1639	1 3/8"	1666	1 5/8"
1640	1 3/8"	1667	1 5/8"
1641	1 3/4"	1668	1 5/8"
1642	1 3/4"	1669	1 5/8"
1643	1 3/4"	1670	1 5/8"
1644	1 3/4"	1671	1 5/8"
1645	1 3/4"	1672	1 5/8"
1646	1 3/4"	1673	1 5/8"
1647	1 3/4"	1674	1 5/8"
1648	1 3/4"		

BOXES

SERIAL NUMBER	AXLE SIZE	SERIAL NUMBER	AXLE SIZE
981	1 3/4"	1000	1 5/8"
982	1 3/4"	1001	1 5/8"
983	1 3/4"	1002	1 5/8"
984	1 3/4"	1003	1 5/8"
985	1 3/4"	1004	1 5/8"
986	1 3/4"	1005	1 5/8"
987	1 3/4"	1006	1 5/8"
988	1 3/4"	1007	1 5/8"
989	1 3/4"	1008	1 1/2"
990	1 3/4"	1009	1 1/2"
991	1 3/4"	1010	1 1/2"
992	1 3/4"	1011	1 1/2"
993	1 5/8"	1012	1 1/2"
994	1 5/8"	1013	1 3/4" 3' X 6" WIDE
995	1 5/8"	1014	1 3/4" 3' X 6" WIDE
996	1 5/8"	1015	1 3/4" 3' X 5 1/2" WIDE
997	1 5/8"	1016	2"
998	1 5/8"	1017	1 3/4"
999	1 5/8"	1018	1 3/4"

Table IV, continued; continues 1917 Boxes.

SERIAL NUMBER	AXLE SIZE	SERIAL NUMBER	AXLE SIZE
1019	1 3/4"	1028	1 3/4" WEISS
1020	1 3/4"	1029	1 5/8" SHAPPELL
1021	1 3/4" REISER	1030	1 1/2" BRICKER
1022	1 3/4" NEIN	1031	1 3/4"
1023	1 3/4"	1032	1 3/4"
1024	1 3/8"	1033	1 3/4"
1025	1 3/8"	1034	1 3/4"
1026	1 3/8"	1035	2"
1027	1 3/8"	1036	SLEIGH BOX

HAY FLATS

726	1 5/8" - 14'	752	1 1/2" - 14'
727	1 3/4" - 14'	753	2" - 18'
728	1 3/4" - 16'	754	2" - 18'
729	1 3/4" - 14'	755	1 3/4" - 16'
730	1 3/4" - 14'	756	1 5/8" - 14'
731	1 5/8" - 14'	757	2" - 18'
732	2" - 18'	758	1 3/4" - 16'
733	2" - 18'	759	1 5/8" - 14'
734	2" - 18'	760	2" - 18'
735	1 7/8" - 16'	761	2" - 18'
736	1 3/4" - 18'	762	2" - 18'
737	1 7/8" - 18'	763	1 3/4" - 14'
738	1 7/8" - 18'	764	2" - 18'
739	2" - 18'	765	2" - 18'
740	2 1/4" - 20'	766	1 3/4" - 14'
741	2 1/4" - 20'	767	2" - 18'
742	2 1/8" - 20'	768	2" - 18'
743	2 1/8" - 18'	769	1 3/4" - 16'
744	1 3/4" - 18'	770	TRUCK 11 1/2'
745	1 5/8" - 14'	771	2" - 18'
746	2" - 18'	772	1 3/4" - 16'
747	2" - 18'	773	1 7/8" - 16'
748	2" - 16'	774	2" - 18'
749	2" - 18'	775	2 1/4" - 18'
750	2" - 17'	776	2 1/8" - 18'
751	2" - 18'		

1918

RUNNING GEARS

1675	2"	1672	2"
1676	2"	1673	2"
1677	2"	1674	2"
1678	2"	1675	2"
1679	2"	1676	2"
1670	2"	1677	2"
1671	2"	1678	2"

Table IV, continued; continues 1918 Running Gears.

SERIAL NUMBER	AXLE SIZE	SERIAL NUMBER	AXLE SIZE
1679	2"	1709	2"
1680	2"	1710	2"
1681	2"	1711	1 3/4"
1682	2"	1712	1 5/8"
1683	1 3/4"	1713	1 5/8"
1684	1 3/4"	1714	1 7/8"
1685	1 3/4"	1715	1 5/8"
1686	1 3/4"	1716	1 3/4"
1687	1 3/4"	1717	1 3/4"
1688	1 3/4"	1718	1 3/4"
1689	1 3/4"	1719	1 3/4"
1690	1 3/4"	1720	1 3/4"
1691	1 3/4"	1721	1 3/4"
1692	1 3/4"	1722	2 1/4"
1693	2"	1723	1 5/8"
1694	2"	1724	1 5/8"
1695	1 5/8"	1725	1 3/4"
1696	1 5/8"	1726	2 1/8"
1697	1 5/8"	1727	2 1/8"
1698	1 5/8"	1728	2 1/8"
1699	1 5/8"	1729	2"
1700	2"	1730	2"
1701	2"	1731	2"
1702	2 1/8"	1732	1 3/4"
1703	2 1/8"	1733	1 3/4"
1704	2 1/8"	1734	1 3/4"
1705	1 3/4"	1735	1 3/4"
1706	2"	1736	1 3/4"
1707	2"	1737	1 3/4"
1708	2"		

BOXES

SERIAL NUMBER	AXLE SIZE	SERIAL NUMBER	AXLE SIZE
1037	1 5/8"	1053	1 3/4"
1038	1 5/8"	1054	1 3/4"
1039	1 5/8"	1055	1 3/4"
1040	1 5/8"	1056	1 3/4"
1041	1 5/8"	1057	1 3/4"
1042	1 5/8"	1058	1 3/4"
1043	1 5/8"	1059	1 3/4"
1044	1 5/8"	1060	1 3/4"
1045	1 5/8"	1061	SLEIGH BOX
1046	1 5/8"	1062	1 3/4"
1047	1 5/8"	1063	1 3/4"
1048	1 5/8"	1064	1 3/4"
1049	1 5/8"	1065	1 3/4"
1050	1 5/8"	1066	1 5/8"
1051	1 3/4"	1067	1 5/8"
1052	1 3/4"	1068	1 5/8"

Table IV, continued; continues 1918 Boxes.

SERIAL NUMBER	AXLE SIZE	SERIAL NUMBER	AXLE SIZE
1069	1 5/8"	1078	1 3/4"
1070	1 5/8"	1079	1 3/4"
1071	1 5/8"	1080	1 3/4"
1072	1 5/8"	1081	1 5/8"
1073	1 3/4"	1082	1 5/8"
1074	1 3/4"	1083	1 5/8"
1075	1 3/4"	1084	1 5/8"
1076	1 3/4"	1085	1 5/8"
1077	1 3/4"		

HAY FLATS

SERIAL NUMBER	AXLE SIZE	SERIAL NUMBER	AXLE SIZE
777	1 7/8" - 16'	811	2" - 18'
778	2" - 18'	812	2" - 18'
779	2" - 18'	813	2" - 18'
780	1 7/8" - 16'	814	2 1/8" - 18'
781	2" - 18'	815	2 1/8" - 18'
782	2 1/8" - 18'	816	2 1/8" - 18'
783	2" - 18'	817	2" - 18'
784	2" - 18'	818	1 5/8" - 14'
785	2" - 16'	819	1 5/8" - 14'
786	1 3/4" - 16'	820	1 3/4" - 16'
787	1 5/8" - 14'	821	1 3/4" - 16'
788	1 5/8" - 14'	822	2" - 18'
789	1 5/8" - 14'	823	2 1/4" - 18'
790	1 5/8" - 14'	824	2 1/8" - 18'
791	2" - 16'	825	1 7/8" - 18'
792	2" - 16'	826	2" - 18'
793	2" - 18'	827	1 3/4" - 16'
794	2" - 16'	828	2" - 18'
795	2" - 18'	829	2" - 18'
796	2 1/8" - 18'	830	2" - 18'
797	1 3/4" - 16'	831	2" - 18'
798	1 3/4" - 16'	832	2" - 16'
799	1 3/4" - 16'	833	1 3/4" - 14'
800	1 1/2" - 14'	834	2 1/8" - 18'
801	1 1/2" - 14'	835	2 1/8" - 18'
802	1 5/8" - 16'	836	
803	1 3/4" - 16'	837	2" - 18'
804	2" - 18'	838	2 1/4" - 18'
805	2" - 18'	839	1 3/4" - 14'
806	2" - 18'	840	2" - 18'
807	2" - 18'	841	2 1/8" - 18'
808	2" - 16'	842	2 1/8" - 18'
809	2" - 18'	843	2" - 18'
810	2" - 18'		

Table IV, continued; begins 1919 Running Gears.

SERIAL NUMBER	AXLE SIZE	SERIAL NUMBER	AXLE SIZE

1919

RUNNING GEARS

SERIAL NUMBER	AXLE SIZE	SERIAL NUMBER	AXLE SIZE
1738	1 5/8"	1784	2"
1739	1 5/8"	1785	2"
1740	1 5/8"	1786	2"
1741	2 1/8"	1787	1 3/8"
1742	1 5/8"	1788	1 3/8"
1743	1 5/8"	1789	1 1/4"
1744	1 3/4"	1790	1 1/4"
1745	1 3/4" P. BALTHASER	1791	1 1/4"
1746	1 3/4"	1792	1 3/4"
1747	1 3/4"	1793	1 3/4"
1748	1 3/4"	1794	2"
1749	2"	1795	1 5/8"
1750	2"	1796	1 5/8"
1751	1 3/4"	1797	1 5/8"
1752	1 3/4"	1798	1 7/8"
1753	1 5/8"	1799	1 7/8"
1754	1 5/8"	1800	1 1/2"
1755	1 5/8"	1801	1 1/2"
1756	1 5/8"	1802	1 1/2"
1757	1 3/4"	1803	1 1/2"
1758	1 3/4"	1804	1 1/2"
1759	1 3/4"	1805	1 3/4"
1760	1 3/4"	1806	1 3/4"
1761	2"	1807	1 3/4"
1762	2"	1808	1 3/4"
1763	2"	1809	1 3/4"
1764	1 3/4"	1810	2"
1765	1 3/4"	1811	2"
1766	1 3/4"	1812	2"
1767	1 3/4"	1813	1 5/8"
1768	1 5/8"	1813	1 5/8"
1769	1 5/8"	1814	1 5/8"
1770	1 5/8"	1815	1 5/8"
1771	1 5/8"	1816	1 5/8"
1772	1 3/8"	1817	1 5/8"
1773	1 7/8"	1818	1 5/8"
1774	1 7/8"	1819	1 5/8"
1775	1 7/8"	1820	1 5/8"
1776	1 7/8"	1821	1 5/8"
1777	2 1/4"	1822	1 5/8"
1778	2"	1823	2"
1779	2"	1824	2"
1780	1 5/8"	1825	2"
1781	1 5/8"	1826	2"
1782	1 5/8"	1827	2"
1783	2"		

Table IV, continued; begins 1919 Boxes.

SERIAL NUMBER	AXLE SIZE	SERIAL NUMBER	AXLE SIZE
BOXES			
1086	1 3/4"	1107	1 5/8"
1087	1 3/4"	1108	1 3/4"
1088	1 3/4"	1109	1 3/4"
1089	1 3/4"	1110	1 5/8"
1090	1 3/4"	1111	1 5/8"
1091	1 3/4"	1112	1 5/8"
1092	1 3/4"	1113	1 5/8"
1093	1 3/4"	1114	1 5/8"
1094	1 3/4"	1115	1 3/8" 2 HORSE
1095	1 3/4"	1116	1 5/8"
1096	1 3/4"	1117	1 1/4"
1097	1 3/4"	1118	1 1/4"
1098	1 5/8"	1119	1 1/4"
1099	1 5/8"	1120	1 3/8"
1100	1 5/8"	1121	1 3/8"
1101	1 5/8"	1122	1 5/8"
1102	1 5/8"	1123	1 5/8"
1103	1 5/8"	1124	1 5/8"
1104	1 5/8"	1125	1 5/8" x 11' LONG
1105	1 5/8"	1126	1 7/8"
1106	1 5/8"	1127	1 7/8"
HAY FLATS			
844	1 3/4" - 14'	865	2" - 18'
845	1 5/8" - 16'	866	2" - 18'
846	1 3/4" - 16'	867	2" - 16'
847	2" - 18'	868	1 3/4" - 16'
848	2 1/8" - 18'	869	1 5/8" - 16'
849	2" - 18'	870	1 3/4" - 16'
850	1 3/4" - 16'	871	2" - 16'
851	1 3/4" - 16'	872	1 3/4" - 16'
852	1 3/4" - 16'	873	1 7/8" - 18'
853	1 5/8" - 14'	874	1 7/8" - 18'
854	1 5/8" - 14'	875	1 7/8" - 16'
855	1 3/4" - 14'	876	2" - 18'
856	1 3/4" - 18'	877	2" - 18'
857	1 3/4" - 16'	878	1 1/2" - 14'
858	2 1/8" - 18'	879	1 5/8" - 14'
859	2 1/4" - 20'	880	2" - 18'
860	2 1/4" - 18'	881	2" - 18'
861	2" - 18'	882	1 3/4" - 16'
862	2" - 18'	883	2" - 18'
863	2" - 18'	884	2" - 18'
864	2" - 18'		

Table IV, continued; begins 1920 Running Gears.

SERIAL NUMBER	AXLE SIZE	SERIAL NUMBER	AXLE SIZE

1920

RUNNING GEARS

SERIAL NUMBER	AXLE SIZE	SERIAL NUMBER	AXLE SIZE
1828	2 1/8"	1851	1 7/8"
1829	2 1/8"	1852	1 5/8"
1830	2 1/8"	1853	1 3/4"
1831	1 3/4"	1854	1 3/8"
1832	1 3/4"	1855	1 1/4"
1833	1 3/4"	1856	1 3/4"
1834	1 3/4"	1857	1 3/4"
1835	1 3/4"	1858	1 3/4"
1836	1 3/4"	1859	1 3/4"
1837	1 3/4"	1860	1 3/4"
1838	1 3/4"	1861	1 3/4"
1839	1 3/4"	1862	1 3/4"
1840	2 1/8"	1863	1 3/4"
1841	1 3/4"	1864	1 7/8"
1842	1 3/4"	1865	1 1/2"
1843	1 3/4"	1866	1 1/2"
1844	1 3/4"	1867	1 1/2"
1845	1 5/8"	1868	2"
1846	1 5/8"	1869	2"
1847	1 5/8"	1870	2"
1848	1 5/8"	1871	2"
1849	2 1/8"	1872	1 3/8"
1850	2 1/8"		

BOXES

SERIAL NUMBER	AXLE SIZE	SERIAL NUMBER	AXLE SIZE
1128	1 3/4"	1148	1 5/8"
1129	1 3/4"	1149	1 5/8"
1130	1 3/4"	1150	1 5/8"
1131	1 3/4"	1151	1 5/8"
1132	1 3/4"	1152	1 1/2"
1133	1 3/4"	1153	1 1/2"
1134	1 3/4"	1154	1 1/2"
1135	1 3/4"	1155	1 1/2"
1136	1 3/4"	1156	1 1/2"
1137	1 3/4"	1157	1 3/4"
1138	1 3/4"	1158	1 3/4"
1139	1 3/4"	1159	1 3/4"
1140	1 5/8"	1160	1 3/4"
1141	1 5/8"	1161	1 3/4"
1142	1 5/8"	1162	1 3/4"
1143	1 5/8"	1163	1 5/8"
1144	1 5/8"	1164	1 5/8"
1145	1 5/8"	1165	1 5/8"
1146	1 5/8"	1166	1 5/8"
1147	1 5/8"	1167	1 5/8"

Table IV, continued; continues 1920 Boxes.

SERIAL NUMBER	AXLE SIZE	SERIAL NUMBER	AXLE SIZE
1168	1 5/8" ELMER OXENRIDER	1177	1 3/4"
1169	1 3/8"	1178	1 3/4"
1170	1 3/4"	1179	1 3/4"
1171	1 7/8"	1181	1 3/4"
1172	1 7/8"	1182	1 3/4"
1173	1 1/2"	1183	1 3/4"
1174	1 1/2"	1184	1 3/4"
1175	1 3/4"	1185	1 3/8"
1176	1 3/4"		

HAY FLATS

885	2" - 18'	902	2" - 18'
886	2" - 18'	903	2" - 18'
887	2 1/8" - 18'	904	2" - 18'
888	1 3/4" - 16'	905	1 7/8" - 16'
889	1 3/4" - 16'	906	1 1/2" - 14'
890	2 1/8" - 18'	907	2" - 18'
891	2" - 18'	908	2 1/8" - 18'
892	2 1/8" - 18'	909	1 5/8" - 14'
893	1 3/4" - 16'	910	2" - 16'
894	1 3/4" - 16'	911	2" - 18'
895	1 3/4" - 16'	912	1 7/8" - 18'
896	2 1/8" - 18'	913	2" - 18'
897	2 1/8" - 18'	914	2 1/8" - 18'
898	2" - 18'	915	1 3/4" - 16'
899	1 1/2" - 14'	916	1 3/4" - 16'
900	1 3/4" - 16'	917	1 3/4" - 14'
901	2 1/8" - 18'	918	1 5/8" - 14'

TRUCK BODIES

1	SCHWARTZ CAB	5	SCHWARTZ CAB
2	SCHWARTZ CAB	6	SCHWARTZ CAB
3	SCHWARTZ CAB	7	SCHWARTZ CAB
4	SCHWARTZ CAB	8	SCHWARTZ CAB

1921

RUNNING GEARS

1873	1 5/8"	1884	1 3/4"
1874	1 5/8"	1885	1 1/4"
1875	1 5/8"	1886	1 1/4"
1876	1 5/8"	1887	1 5/8"
1877	2 1/8"	1888	1 5/8" R. F. MOYER
1878	2 1/8"	1889	1 5/8"
1879	2 1/8"	1890	1 5/8"
1880	1 3/4"	1891	1 5/8"
1881	1 3/4"	1892	1 5/8"
1882	1 3/4"	1893	1 5/8"
1883	1 3/4"	1894	1 5/8"

Table IV, continued; continues 1921 Running Gears.

SERIAL NUMBER	AXLE SIZE	SERIAL NUMBER	AXLE SIZE
1895	1 5/8"	1902	1 1/2"
1896	1 7/8"	1903	1 1/2"
1897	1 7/8"	1904	1 1/2"
1898	1 7/8"	1905	2"
1899	1 3/4"	1906	1 3/4"
1900	1 3/4"	1907	1 3/4"
1901	1 3/4"	1908	1 3/4"

BOXES

1186	1 1/4"	1195	1 1/2"
1187	1 1/4"	1196	1 1/2"
1188	1 3/4"	1197	1 3/4"
1189	1 7/8"	1198	1 3/4"
1190	1 1/2"	1199	1 3/4"
1191	1 1/2"	1200	1 3/4"
1192	1 1/2"	1201	1 3/4"
1193	1 1/2"	1202	1 3/4"
1194	1 1/2" LYMAN FIDLER		

HAY FLATS

919	2" - 18'	928	1 3/4" - 16'
920	2" - 18'	929	1 3/4" - 16'
921	1 5/8" - 14'	930	2" - 16'
922	1 3/4" - 16'	931	2 1/8" - 18'
923	2" - 18'	932	2" - 18'
924	2" - 18'	933	2" - 18'
925	2" - 18'	934	2 1/8" - 18'
926	2" - 18'	935	2" - 18'
927	2" - 18'		

TRUCK BODIES

9	IHC CAB	19	EXPRESS BODY
10	BODY ROUND TOP	20	EXPRESS BODY
11	BODY ROUND TOP	21	PANEL BODY BEER
12	BODY ROUND TOP	22	BOX BODY
13	BODY ROUND TOP	23	BAKER BODY
14	BODY ROUND TOP	24	MEINIG
15	BODY ROUND TOP	25	PASSENGER BUS
16	BODY ROUND TOP	26	PASSENGER BUS
17	BODY ROUND TOP	11 PCS	TRUCK SEATS
18	BODY ROUND TOP		

1922

RUNNING GEARS

1909	1 5/8"	1913	1 3/4"
1910	1 5/8"	1914	1 3/4"
1911	1 5/8"	1915	1 3/4"
1912	1 3/4"	1916	1 3/4"

Table IV, continued; continues 1922 Running Gears.

SERIAL NUMBER	AXLE SIZE	SERIAL NUMBER	AXLE SIZE
1917	1 3/4"	1922	1 5/8"
1918	1 3/4"	1923	1 5/8"
1919	1 1/4"	1924	1 5/8"
1920	1 1/4"	1925	1 1/4"
1921	1 1/4"	1926	2 1/2" BULL WAGON

BOXES

1203	1 5/8"	1210	1 5/8"
1204	1 5/8"	1211	1 5/8"
1205	1 5/8"	1212	1 5/8"
1206	1 5/8"	1213	1 5/8"
1207	1 5/8"	1214	1 5/8"
1208	1 5/8"	1215	?
1209	1 5/8"		

HAY FLATS

936	1 3/4" - 16'	942	2" - 16'
937	1 3/4" - 16'	943	1 3/4" - 14'
938	2" - 18'	944	1 3/4" - 16'
939	1 5/8" - 14'	945	1 3/4" - 14'
940	2" - 18'	946	1 3/4" - 16'
941	1 5/8" - 14'	947	2" - 18'

TRUCK BODIES

27	WITMAN	45	MOUNTZ REBUILD
28	FRUIT TRUCK	46	SCHNECK BROS. REBUILD
29	TRUCK BODY W/ TOP	47	LUTZ
30	BODY & CAB	48	PAPER MILL
31	BODY & CAB	49	OPEN CAB, BODY LOOSE
32	TELEPHONE BODY		POTTSVILLE
33	COAL & LUMBER BODY	50	BAKER BODY MAIER
34	EXPRESS	51	1 TON CLOSED CAB, DROP
35	BAKER BODY		BACK, POTTSVILLE
36	MILK BODY	52	KREITZ FLAT
37	MILK BODY	53	WHITE TRUCK JAMES
38	PIANO BODY		FURNITURE & MOVING
39	PRETZEL BODY	54	WHITE TRUCK JAMES
40	DYE BODY		FURNITURE & MOVING
41	POTTSVILLE	55	BUTCHER BODY,
42	POTTSVILLE		SCHWARTZ
43	FAIR	56	WHITE TRUCK JAMES
44	OPEN EXPRESS		FURNITURE & MOVING

Table IV, continued; begins 1923 Running Gears.

SERIAL NUMBER	AXLE SIZE	SERIAL NUMBER	AXLE SIZE

1923

RUNNING GEARS

1927	1 3/8"	1930	1 3/4"
1928	1 3/8"	1931	1 3/4" LOW
1929	1 3/8"	1932	1 3/4" LOW

BOXES

1216	1 1/4"	1222	1 3/4"
1217	1 1/4"	1223	1 3/4"
1218	1 3/8"	1224	1 3/4" JOHN EARLY
1219	1 3/8"	1225	1 3/4" ALVIN KNOLL
1220	1 3/8"	1226	1 3/4"
1221	1 3/8"		

HAY FLATS

948	2" - 18'	954	2" - 18'
949	1 3/4" - 16' STANFORD	955	1 5/8" - 14'
	LANDIS	956	1 3/4" - 16'
950	1 3/4" - 16'	957	1 3/4" - 16'
951	2" - 18'	958	2 1/4" - 18'
952	2 1/8" - 18'	959	2" - 18'
953	2 1/8" - 18'		

TRUCK BODIES

57	BODY & CAB SCHWARTZ	74	GMC PANEL BODY
58	BODY 18'	75	PIERCE ARROW BODY
59	CAB	76	SCHWARTZ CAB & BODY
60	BODY SPANNUTH	77	GMC
61	BODY ROUND TOP	78	MASON TRUCK BODY
62	SCHWARTZ CAB	79	AUTO CAR FLAT
63	1 TON OPEN EXPRESS	80	STAKE FLAT
64	GMC STAKE BODY	81	SCHWARTZ CAB & BODY
65	SCHWARTZ CAB	82	WHITE BODY & CAB
66	BAKER BODY	83	DUMP BODY
67	SCHWARTZ CAB	84	WHITE BODY & CAB
68	SCHWARTZ CAB	85	CLINTON TRUCK VAN
69	BODY & CAB		BODY
70	STAKE BODY	86	PRETZEL BODY
71	5 TON BODY	87	FURNITURE BODY
72	2 TON TOP	88	FURNITURE BODY
73	AUTO CAR FLAT		

Table IV, continued; begins 1924 Running Gears.

SERIAL NUMBER	AXLE SIZE	SERIAL NUMBER	AXLE SIZE

1924

RUNNING GEARS

1933	1 1/2"	1938	1 5/8"
1934	1 1/2"	1939	2"
1935	1 5/8"	1940	2"
1936	1 5/8"	1941	1 3/8" TOP SPRING
1937	1 5/8" ELMER OXENRIDER		PLATFORM

HAY FLATS

960	1 3/4" - 16'	964	1 3/4" - 16'
961	1 5/8" - 14'	965	2" - 18'
962	2" - 18'	966	1 5/8" - 14'
963	1 5/8" - 14' READING FAIR	967	1 3/4" - 16'

TRUCK BODIES

89	PRETZEL BODY	98	CAB
90	EXPRESS BODY	99	BODY
91	MOVING VAN PIERCE ARROW	100	BAKER PANEL BODY
		101	OPEN EXPRESS BODY
92	PANEL BODY AUBURN BODY	102	FURNITURE BODY
		103	FRUIT TRUCK BODY
93	FRUIT TRUCK BODY	104	RUNABOUT BODY
94	GMC CAB BECHTOLD	105	INTERNATIONAL EXPRESS BODY
95	INTERNATIONAL TRUCK CAB	106	FURNITURE BODY
96	INTERNATIONAL TRUCK CAB	107	FURNITURE BODY
		108	INTERNATIONAL BUS BODY
97	EXPRESS BODY CLINTON		

1925

RUNNING GEARS

1942	1 3/4"	1949	1 1/2" 1 HORSE WAGON STEFFY
1943	1 3/4" LYMAN ELDLER		
1944	1 1/4" WM MARKS	1950	2"
1945	1 1/4"	1951	1 3/4"
1946	1 3/8"	1952	1 3/4"
1947	1 3/8"	1953	1 3/4"
1948	1 3/8"	1954	1 3/4"
		1955	1 7/8" WALLACE NOLL

BOXES

1227	1 1/4"	1230	1 3/8"
1228	1 1/4"	1231	1 3/8"
1229	1 3/8"	1232	1 3/4" JAMES SCHNADER

Table IV, continued; begins 1925 Hay Flats.

SERIAL NUMBER	AXLE SIZE	SERIAL NUMBER	AXLE SIZE

HAY FLATS

968	2" - 18' BUCKLES	972	1 3/4" - 16'
969	2" - 18' END GATE, BUCKLES, HASSLER	973	2" - 16'
		974	2" - 18'
970	2" - 18'	975	2" - 18' MILES KALBACH
971	2 1/4" - 20'	976	1 3/4" - 16' LOW WHEELS W. NOLL

TRUCK BODIES

109	PANEL BODY FORD	115	STAKE FLAT BED
110	OPEN EXPRESS	116	CAB CLYDESDALE
111	CAB AUTO CAR	117	EXPRESS BODY
112	PANEL BODY AUTO TRUCK	118	PLUMBING BODY ON WHITE
113	FORD SLIP ON BODY		
114	FURNITURE CLINTON BODY	119	HOSIERY MILL ON JORDAN

1926

RUNNING GEARS

1956	1 3/4" ALVIN KNOLL	1964	1 3/4" LOW, ELMER HOFFMAN
1957	1 3/4" READING FAIR		
1958	1 3/4" J. W. STETZLER	1965	1 3/4" LOW, ELMER HOFFMAN
1959	1 3/4" CHAS SHEIDY		
1960	1 3/4" ADAM ABMACHT	1966	1 3/4" WEBBER
1961	1 3/4" SAM HALDAMAN	1967	1 3/4" WEBBER
1962	1 3/4"	1968	1 5/8" - 2 1/2" TIRES, WALBORN
1963	1 3/4" JAMES SCHNADER		

BOXES

1233	1 3/4" WEBBER SOILD BODY	1234	1 3/4" WEBBER SOLID BODY

HAY FLATS

977	1 3/4" - 14'	984	2" - 18' DAVIS BURGERT
978	1 3/4" - 14'	985	2" - 18'
979	1 3/4" - 14'	986	16' LOW HOFFMAN
980	1 3/4" - 14' OPEN END HAFER	987	16' PIERCE LESHER, STEP SEAT & BRAKE
981	1 3/4" - 14'	988	JOHN DIETRICH
982	J.W. STETZLER STEP,SEAT,BRAKE,RODS	989	18' LOW, MORRIS ROTHENBERGER
983	ADAM ABMACHT		

TRUCK BODIES

120	PRETZEL SLIP ON	121	PENN BOTTLING WORKS

Table IV, continued; begins 1927 Running Gears.

SERIAL NUMBER	AXLE SIZE	SERIAL NUMBER	AXLE SIZE

1927

RUNNING GEARS

SERIAL NUMBER	AXLE SIZE	SERIAL NUMBER	AXLE SIZE
1969	1 5/8" - 2 1/2" TIRES, LOW	1978	1 3/4" - 3" TIRES, LOW,
1970	1 5/8" - 2 1/2" TIRES, LOW,		ELMER GOCKLEY
	ISSAC EBERLY	1979	1 3/4" - 3" TIRES, LOW,
1971	1 3/4" - 3" TIRES, LOW,		ELMER GOCKLEY
	ALBERT ARNOLD	1980	1 3/4" - 3" TIRES, LOW
1972	1 3/4" - 3" TIRES, LOW,	1981	1 5/8" - 2 1/2" TIRES, LOW,
	ALBERT ARNOLD		C.W. HOFFER
1973	1 3/4" - 3" TIRES, LOW,	1982	1 3/4" - 3" TIRES, LOW,
	JAMES KLOSE		PHARES KRALL
1974	1 3/4" - 3" TIRES, LOW, JOHN	1983	1 5/8" - 2 1/2" TIRES, LOW,
	RUTH		JOHN SCHOCK
1975	1 7/8" - 2 1/2" TIRES, LOW,	1984	1 5/8" - 2 1/2" TIRES, LOW,
	DR HAIN		IRWIN FIDLER
1976	1 7/8" - 2 1/2" TIRES, LOW,	1985	1 3/4" - 3" TIRES, LOW
	DR HAIN	1986	1 3/4" - 3" TIRES, LOW
1977	1 3/4" - 3" TIRES, LOW	1987	1 3/4" - 3" TIRES, LOW,
	MONROE SLAK		WESLEY GRAYBILL

HAY FLATS

SERIAL NUMBER	AXLE SIZE	SERIAL NUMBER	AXLE SIZE
990	18' LOW, MORRIS	1005	18' CHAS SPEICHER
	ROTHENBERGER	1006	1 5/8" - 2 1/2" TIRES, LOW,
991	18' LOW, HOWARD SNYDER		14' JOHN SCHOCK
992		1007	16' TROUTMAN
993		1008	
994		1009	
995	1 3/4" - 3" TIRES, LOW, 16'	1010	16' ISSAC EBERLY
	ALBERT ARNOLD	1011	16' CHAS HAIN
996	1 3/4" - 3" TIRES, LOW, 16'	1012	16' SEAT & BRAKE PHARES
	ALBERT ARNOLD		KRALL
997	16' LOW, ELMER GOCKLEY	1013	16' SEAT & BRAKE JOHN
999	1 7/8" - 2 1/2" TIRES, LOW,		WIENER
	18' DR HAIN	1014	14' LOW WM SNYDER
1000	1 3/4" - 3" TIRES, LOW,18'		(FAIR)
	JAMES KLOSE	1015	2" - 18' JAMES HERB
1001	18' CHARLES KAUFFMAN	1016	16' LOW, (FAIR)
1002	16' CAL SHIEDY	1017	2" - 18'
1003	18'	1018	2" - 18'
1004	18' WARREN		
	HIMMELBERGER		

Table IV, continued; begins 1928 Running Gears.

SERIAL NUMBER	AXLE SIZE	SERIAL NUMBER	AXLE SIZE

1928

RUNNING GEARS

SERIAL NUMBER	AXLE SIZE	SERIAL NUMBER	AXLE SIZE
1987	1 3/4" - 3" TIRES, LOW, U.L. GRAEFF	1998	1 3/4" - 3" TIRES, LOW,
1988	1 3/4" - 3" TIRES, LOW, WERLEY GRAYBILL	1999	1 3/4" - 3" TIRES, LOW, RUDY FREDERICKSBURG
1989	1 3/4" - 3" TIRES, LOW,	2000	1 3/4" - 3" TIRES, LOW, ISSAC PHILLIPS
1990	1 3/4" - 3" TIRES, LOW, REBUILD ENGINE WAGON HOWARD SCHLAPPICH	2001	1 3/4" - 3" TIRES, LOW,
		2002	1 3/4" - 3" TIRES, LOW, BERKS CO PRISON
1991	1 5/8" - 2 1/2" TIRES, LOW,	2003	
1992	1 5/8" - 2 1/2" TIRES, LOW,	2004	1 1/2" - 3" TIRES, LOW, SPRING BOLSTERS, WERNERSVILLE STATE HOSP.
1993	1 3/4" - 3" TIRES, LOW, U.L.GRAEFF		
1994	1 3/4" - 3" TIRES, LOW,	2005	11/16 18"- 22" WHEELS HANDWAGON ED HAFER
1995	1 3/4" - 3" TIRES, LOW,		
1996	1 3/4" - 3" TIRES, LOW, RUDY FREDERICKSBURG	2006	1 3/4" - 3" TIRES, LOW
1997	1 3/4" - 3" TIRES, LOW,	2007	1 3/4" - 3" TIRES, LOW

HAY FLATS

SERIAL NUMBER	AXLE SIZE	SERIAL NUMBER	AXLE SIZE
1019	16' OSCAR MANBECK	1029	16'
1020	16'	1030	16'
1021	16' U.L. GRAEFF	1031	16' BALTHASER
1022	16'	1032	14' MATZ
1023	16'	1033	16' OLD WAGON KAUFFMAN
1024	16' HOWARD SCHLAPPICH		
1025	18' WARREN MOLL	1034	14' KERCHNER
1026	14' LOW WHEEL A.B. MADEIRA	1035	16' LOW WHEEL KIRKOFF
		1036	18' 3 TON, AARON KOCH
1027	16' LOW WHEEL U.L. GRAEFF	1037	18' WERNERSVILLE STATE HOSP.STEP, SEAT, BRAKE
1028	16'	1038	18' STATE HILL FRUIT FARM

TRUCK BODIES

SERIAL NUMBER	AXLE SIZE
122	HERBERT SHEIDY

1929

RUNNING GEARS

SERIAL NUMBER	AXLE SIZE	SERIAL NUMBER	AXLE SIZE
2008	11/16" HAND WAGON	2011	1 3/4" - 3" TIRES, LOW, OSCAR MANBECK
2009	11/16" HAND WAGON GEO BARE	2012	1 5/8" - 2 1/2" TIRES, LOW
2010	1 3/4" - 3 5/8" TIRES, LOW, FRANK RUPPERT	2013	1 5/8" - 2 1/2" TIRES, LOW, ST FRANCIS ORPHANAGE

Table IV, continued; begins 1929 Hay Flats.

SERIAL NUMBER	AXLE SIZE	SERIAL NUMBER	AXLE SIZE
	HAY FLATS		
1039	16'	1044	16' FAIR WAGON SEIFERT
1040	16'	1045	14'
1041	16' COUNTY COMMISSIONERS	1046	16' POTATO FLAT STOLTZFUS
1042	HAND WAGON GEORGE BARE	1047	16' TOBACCO RACK STOLTZFUS
1043	16' SCHAPPEL		
	TRUCK BODIES		
123	FORD PANEL BODY	128	8' W x 13' L AUTO CAR OPEN BODY ISSAC EBERLY
124	FORD SLIP ON BODY ADAM GRETH		
125	FORD EXPRESS REMOVABLE TOP	129	6' L CHEVROLET SLIP ON
		130	WIPPET EXPRESS JOHN NEFF ROBESONIA
126	STAKE BODY PIERCE ARROW WEAVER	131	DODGE COUPE SLIP ON BODY SAMUEL SIMMON
127	FORD SLIP ON BODY JOHN RICK	132	PAUL DEGLER LOOSE SIDES 12' L

1930

RUNNING GEARS

SERIAL NUMBER	AXLE SIZE	SERIAL NUMBER	AXLE SIZE
2014	2 1/8" - 4" TIRES, LOW, WERNERSVILLE STATE HOSP.	2017	1 3/4", 3" TIRES, LOW, OSCAR SPAYD
		2018	1 3/4", 3" TIRES, LOW, ISSAC EBERLY
2015	1 3/4" - 3" TIRES, LOW, JOHN KALBACH	2019	1 5/8", 2 1/2" TIRES, LOW
2016		2020	WHEELS 24" x 30" H. M. FOCHT

HAY FLATS

SERIAL NUMBER	AXLE SIZE	SERIAL NUMBER	AXLE SIZE
1048	16'	1062	14' BERKS CO PRISON JUNE 1935
1049	16' JOHN RUTH		
1050	16' ISSAC PHILLIPS	1063	16' ISAAC EBERLY
1051	16' BERKS CO BERKS	1064	16' ST FRANCIS ORPHANAGE
1052	16'		
1053	16'	1065	16' MILTON BOLTZ
1054	18'	1066	18' HERBERT REEDY
1055	14'	1067	14' ROTHERMOL
1056	18' W/RODS WERNERSVILLE STATE HOSPITAL ON 2 1/8" GEAR	1068	16' FLAT BED H. M. FOCHT
		1069	2" - 18'
		1070	2" - 18' JAMES FEICK
		1071	17' IRON WAGON, ORCHARD BED, STATE
1057	16'		HILL FRUIT FARM
1058	16' GEORGE WAGNER		
1059	18'	1072	17' IRON WAGON, ORCHARD BED, STATE
1060			HILL FRUIT FARM
1061	1 3/4 - 16' WM BRIGHT JUNE 1934		

Table IV, continued; begins 1930 Truck Bodies.

SERIAL NUMBER	AXLE SIZE	SERIAL NUMBER	AXLE SIZE
	TRUCK BODIES		
133	EXPRESS BODY REMOVABLE TOP, HINKLEMAN &SONS	135	DODGE COUPE SLIP ON BODY CHARLES HECK
134	OPEN EXPRESS BODY SAMUEL WEIDENHAMMER	136	CHEVY TRUCK, HARRY HOLLENBACH
		137	FARM DELIVERY BODY, CATTLE RACK, CHAS REIGEL

1931

RUNNING GEARS

2021	1 5/8" - 3" TIRES x 1/2", LOW	2026	1 3/4" - 3" TIRES x 5/8", LOW, WHLS, 2' 3" x 2' 10" JOHN STOLTZFUS
2022	1 5/8" - 2 1/2" TIRES x 5/8", LOW		
2023	1 3/4" - 3" TIRES x 5/8", LOW JOHN RUTH	2027	1 3/4" - 3" TIRES x 5/8", LOW, WHLS, 24" x 30"
2024	1 3/4" - 3" TIRES x 5/8", LOW MILTON BOLTZ	2028	1 3/4" - 3" TIRES x 5/8", LOW, WHLS 2' 7" x 3' 2", CLAYTON TOBIAS
2025	1 3/4" - 3" TIRES x 5/8", LOW, WHLS 2' 7" x 3' 2"		

TRUCK BODIES

138	HOWARD KISSINGER

1932–1933 (no records)

1934

HAY FLATS

1073	18' FRANK RUPPERT AUGUST 1934	1076	16' FLAT BED H. M. FOCHT
		1077	18' WM WEIDMAN
1074	18' - HEAVY, CHAS KEENER	1078	16' CLAYTON TOBIAS
1075	16' FLAT BED H. M. FOCHT	1079	18' W/RODS JOHN RUTH

Present Location of Some Gruber Wagons

Berks County Parks & Recreation Department at Berks County Heritage Center (Not a complete listing)

Running gear #24 built in1904 for William Reber
Running gear #46 built in 1904 for A. L. Roth
Running gear #116 built in 1906
Running gear #376 built in1908 for John DeTurk
Running gear #608 built in1908
Running gear #1468 built in 1915
Running gear #1528 built in 1916
Running gear #1618 built in 1917
Running gear #1648 built in 1917 for Chas. Heck
Running gear #1657 built in 1917 for Samuel Schell
Running gear #1908 built in 1921 for Ephrata Body
Running gear #1940 built in 1924
Running gear (number not found)
Wagon box #328 built in 1908 for Cyrus Ruth
Wagon box #425 built in 1908 for Levi Spatz
Wagon box #974 built in 1916
Hay flat #73 built in 1905 for Samuel Koller
Hay flat #599 built in 1914
Hay flat #665 built in 1915
Hay flat #845 built in 1919 for Paul Balthaser
Hay flat #1015 built in 1927 for James Herb
Hay flat #1021 built in 1928 for U. L. Graeff
Hay flat #1062 built in 1935 for Berks County Prison

Boyertown Museum of Historic Vehicles, Boyertown, PA

Running gear #1643 built in 1917
Hay flat #772 built in 1916

Carriage Museum, The Museums at Stony Brook, Stony Brook, NY

Running gear (number unknown)
Hay flat (number unknown)

Deisemann, Rob, Bernville, PA

Running gear #202 built in 1905 for Frank Eckenroth
Running gear #1509 built in 1916
Running gear #1588 built in 1916

Deisemann, Rob, continued

> Running gear #1910 built in 1922
> Wagon box #1069 built in 1918
> Hay flat #74 built in 1905 for Frank Eckenroth
> Hay flat #184 built in 1907 for Levi Becker
> Hay flat #1038 built in 1928 for State Hill Fruit Farms

Etchberger, Arthur and Catherine, Bernville, PA

> Running gear #11 built in 1904 for George Beahl
> Running gear #605 built in 1909 for Samuel Startzer
> Running gear (One Horse) #1946 built in 1925
> Wagon box (number not found)
> Hay flat #18 built in 1904 for John A. Rigg
> Hay flat #137 built in 1906 for William Forry
> Hay flat #626 built in 1915

Hans Herr House, Willow Street, PA

> Running gear (number not found)
> Hay flat #1065 built in 1931 for Milton Boltz

Henry Ford Museum and Greenfield Village, Dearborn, MI

> Running gear (number not found)
> Hay flat #391 built in 1911 for Isaac Speicher

Morey, Susan, Mohnton, PA

> Running gear #1671 built in 1917
> Hay flat #726 built in 1917

Nagle, Dennis, Reading, PA

> Running gear #1562 built in 1916
> Hay flat #719 built in 1916

Ray, Clayton, Falmouth, VA

> Running gear #1633 built in 1917
> Running gear #1992 built in 1928
> Hay flat #756 built in 1917
> Hay flat #1031 built in 1928 for Balthaser

Schrack, Charles, Bernville, PA

> Running gear #45 built in 1904
> Hay flat #81 built in 1905

Swope, Edward, Bernville, PA

  Running gear #75 built in 1904 for Reading Sanitary R
    Company
  Running gear #1153 built in 1913
  Running gear #1254 built in 1913
  Running gear #1687 built in 1918 (with hay flat)
  Running gear #1925 built in 1922 (now box with original
    iron work)
  Running gear (number not found)
  Wagon box #466 built in 1909 for Edward Filbert
  Wagon box #848 built in 1914
  Hay flat (number not found)
  Hay flat (number not found)

Wentzel, Tom and Susan, Wyomissing, PA

  Running gear #1082 built in 1912

Literature Cited

Anonymous
  1972   West's patent hydraulic tyre setter. *Carriage Journal*, 9(4):184–186, 3 unnumbered figures. (Reprinted from *The Coach Builders', Harness Makers' and Saddlers' Art Journal*, 15 March 1895.)

  1974   Wagons for farmers in south New Jersey. *Carriage Journal*, 11(4):177–178, 2 unnumbered figures. (H. A. DeHart, Thorofare, New Jersey.)

Arnold, James
  1969   *The Farm Waggons of England and Wales.* John Baker, London, 23 + [25] pages, 24 plates, 5 unnumbered text figures, drawings on endpapers, duplicated front and back. (Reprinted with corrections, 1974; reprinted, 1978.)

Bailey, Jocelyn
  1975   The village wheelwright. *Shire Album* 11, 32 pages, unnumbered illustrations. (Published by Shire Publications Ltd., Aylesbury, Bucks, United Kingdom; reprinted and somewhat modified at least through 1989.)

Bailey, Robbie
  1992   The Zuraw wagon. *Foxfire Magazine*, Fall–Winter 1992:187–188, 2 unnumbered figures.

Berkebile, Don H.
  1959   Conestoga wagons in Braddock's Campaign, 1755. *Smithsonian Institution, United States National Museum, Bulletin 218, Contributions from the Museum of History and Technology, Paper* 9:141–154, 10 figures.

  1978   *Carriage Terminology: An Historical Dictionary.* Smithsonian Institution Press, Washington, DC (copublished with Liberty Cap Books), 488 pages, numerous unnumbered illustrations.

Boyertown Times
   1965   Sailor transfers teaching from Navy ships to 'shop.'
           Boyertown Times, 108(48):1,8, one unnumbered figure (April
           29, 1965).

Bristol Wagon & Carriage Works
   1994   *Bristol Wagon & Carriage Illustrated Catalog, 1900*. [viii] &
           178 pages, numerous unnumbered illustrations. (Originally
           published 1900; republished 1994 by Dover Publications,
           Mineola, New York.)

Budd, Thomas, and Claude L. Brock, editors
   1996   *Farming Once Upon a Time; More Remarkable Photos by J. C.
           Allen & Son, 1912–1952*. Concord Publishers, Louisville,
           Kentucky. 160 pages, and more than that many
           unnumbered photos. (A similar collection of the Allens'
           photos was published in 1995 by the same people and
           others, but is not cited in this one, although implied by the
           title.)

Cannon, William A., and Fred K. Fox
   1981   *Studebaker: The Complete Story*. Tab Books Inc., Blue Ridge
           Summit, Pennsylvania, 368 pages, many unnumbered
           illustrations.

Carter, Edward C., II, John C. Van Horne, and Charles E. Brownell, editors
   1985   *Latrobe's View of America, 1795–1820: Selections from the
           Watercolors and Sketches*. Published for the Maryland
           Historical Society by Yale University Press, New Haven,
           Connecticut, xxii + 400 pages, 18 figures, 161 plates, 4 maps.

Cech, Daniel J.
   1982   Gruber Wagon Works celebrates its 100th anniversary.
           Carriage Journal, 19(4):80–81, 4 unnumbered figures.

Chadsey, Charles P., William Morris, and Harold Wentworth, editors
   1947   *Words: The New Dictionary*. Grosset & Dunlap, New York,
           xxii + 704 pages, many unnumbered figures.

Columbia Wagons
   1917   Illustrated catalogue of Columbian Line wagons and carts.
           Heavy wagons, no. 36. Columbia Wagon Co., Columbia,
           Pennsylvania, 73 pages, many unnumbered illustrations.
           (Page 4 alludes to this as "our new Catalog No. 37.")

Cook, Charles W.

2005   Moving the Gruber Wagon Works. Video tape produced by Charles W. Cook of R. S. Cook and Associates, Inc., Southeastern, Pennsylvania, which was the general contractor for the move in 1976.

Cooper, Carolyn C.

1994   A patent transformation: Woodworking mechanization in Philadelphia, 1830–1856. Pages 278–327, 16 figures, *in* Judith A. McGaw, editor, *Early American Technology: Making and Doing Things from the Colonial Era to 1850*. University of North Carolina Press, Chapel Hill, North Carolina (published for the Institute of Early American History and Culture, Williamsburg, Virginia).

Cope, Kenneth L.

2004   *Carriage and Wagon Makers' Machinery and Tools*. Astragal Press, Mendham, New Jersey, 200 pages, many unnumbered illustrations.

Cousins, Peter H.

1990   *Old-fashioned Farm Life Coloring Book: Nineteenth Century Activities on the Firestone Farm at Greenfield Village*. Dover Publications, Mineola, New York, [iv] + 44 pages, many unnumbered illustrations by A. G. Smith.

Critchlow, Donald T.

1996   *Studebaker: The Life and Death of an American Corporation*. Indiana University Press, Bloomington, Indiana, xiv + 273 pages, some unnumbered illustrations.

Daniel, J. M., Jr.

1979   *Hackney, the History of a Company*. Hackney Brothers Body Company, Wilson, North Carolina. [iv] + xii + 117 pages, 18 unnumbered figures, 16 of which are portraits of Hackney family members.

Dewitt, Melvin L.

1984   *Wheels, Wheels, Wagons & More*. Pine Creek Industries, Pinehurst, Idaho, [ii] + ii + 106 pages, many unnumbered illustrations.

Dunbar, Seymour

1937   *A History of Travel in America*. New edition, Tudor Publishing Company, New York, liv + 1531 pages, 2 maps, 12 colored plates, 400 illustrations (first edition 1915, Bobbs-Merrill Company, Indianapolis, Indiana).

Egerton, Douglas R.
   1993   An upright man: Gabriel's Virginia and the path to slave
           rebellion. *Virginia Cavalcade*, 43(2):52–69, 15 unnumbered
           figures.

Ellis, Franklin, and Samuel Evans
   1883   *History of Lancaster County, Pennsylvania, With Biographical
           Sketches of Many of Its Pioneers and Prominent Men.* Everts &
           Peck, Philadelphia, 1101 pages, illustrated.

Erskine, Albert Russell
   1924   *History of the Studebaker Corporation.* Studebaker
           Corporation, South Bend, Indiana, xxvii + 229 pages,
           frontispiece, many unnumbered illustrations.

Eshelman, Ralph E., Robert J. Emry, Daryl P. Domning, and David J. Bohaska
   2002   Biography and bibliography of Clayton Edward Ray. Pages
           1–13, frontispiece, Figures 1–3, *in* Robert J. Emry, editor,
           Cenozoic mammals of land and sea: tributes to the career
           of Clayton E. Ray. *Smithsonian Contributions to Paleobiology*,
           93:vi + 372 pages, many illustrations.

Farrow, Edward S.
   1895   *Farrow's Military Encyclopedia: A Dictionary of Military
           Knowledge, Illustrated With Maps and About Three Thousand
           Wood Engravings.* Military-Naval Publishing Company,
           New York, second edition, volume I, 821 pages, plus
           frontispiece and 14 pages of illustrations following page
           821; many illustrations, in part unnumbered and in part
           numbered separately for each entry.

Fegley, H. Winslow
   1916   A family of successful blacksmiths and wheelwrights. *The
           American Blacksmith: A Practical Journal of Blacksmithing and
           Wagonmaking*, 15(4):86–89, 8 unnumbered illustrations.

Florin, Lambert
   1970   *Western Wagon Wheels. A Pictorial Memorial to the Wheels
           That Won the West.* Superior Publishing Company, Seattle,
           183 pages, many unnumbered photographs.

Frizzell, John, and Mildred Frizzell
   1977   In quest of a Coquillard wagon. *Carriage Journal*, 15(3):358–
           360, 3 unnumbered figures. (Coquillard Wagon Company,
           South Bend, Indiana.)

Green, Susan

    2002   *Hitch Wagons for City Driving & More*. Carriage Museum of America, Bird-in-Hand, Pennsylvania, 392 pages, many unnumbered illustrations. (Year of publication not indicated in the book, but confirmed by Susan Green, Project Director for the book, personal communication, 11 August 2005.)

    2005   *Pennsylvania Carriage and Wagon Makers*. http://www.carriagemuseumlibrary.org/penn_carriage_makers.htm. (The bulk of the list is from Wilson H. Lee, *American Carriage Directory*, 1898, Volume ll, Price and Lee Co., New Haven, Connecticut, to which additions are made as found; Green, personal communication, 16 May 2005.)

Hafer, Erminie Shaeffer

    1972   *A Century of Vehicle Craftsmanship: A One Hundred Year History of the Pennsylvania Dutch Transportation Heritage of Berks County: Bicycles, Wagons, Buggies, Sleighs, Carriages, Fire Engines, Steam Trains, Trolley Cars, Buses, Trucks, and Automobiles*. Hafer Foundation, Boyertown Museum of Historic Vehicles, Boyertown, Pennsylvania, 264 pages, many unnumbered illustrations.

Hendrikson, M. C.

    1997   *The Secrets of Wheelwrighting: Tyres*. M. C. & P. Hendrikson, Kariong, New South Wales, [ii] + 91 pages, many unnumbered illustrations. (Especially good on West's cold "tyre" setter and on strakes.)

Homme, Ferd

    1947   *Oak Opening: The Story of Stoughton*. Stoughton Centennial History Committee, Stoughton, Wisconsin, 82 pages, illustrated.

Hughes, Ralph C.

    1995   *John Deere Buggies and Wagons*. American Society of Agricultural Engineers, St. Joseph, Michigan. viii + 64 pages, many unnumbered illustrations. (Part II: wagons, pages 16–24, 28–54, 56–64.)

Hunsberger, Bruce

  In Press The Gruber Body Works. *The Historical Review of Berks County*.

Hunsberger, Carol
   2005   *The Gruber Wagon Works: The Place Where Time Stood Still.*
           The Society for the Preservation of the Gruber Wagon
           Works, Leesport, Pennsylvania.

Hutchins, Daniel D.
   2004   *Wheels Across America: Carriage Art & Craftsmanship.* Tempo
           International Publishing Company, Santa Fe, New Mexico,
           vi + 234 pages, many unnumbered illustrations. (Identified
           as volume 1 on title page; conceived as a multi-volume
           series, pages iv and 230.)

Jenkins, J. Geraint
   1972   *The English Farm Wagon: Origins and Structure.* Second
           edition. David and Charles, Newton Abbot, xii + 250 pages,
           58 illustrations. (First edition, 1961.)

John Milner Associates
   1978   Gruber Wagon Works, Blue Marsh Lake, Berks County,
           Pennsylvania. Historic Structures Report, prepared for US
           Army Engineer District Philadelphia, Corps of Engineers,
           US Army, Philadelphia, Pennsylvania 19106, Contract
           DAWCW61-78-C-0038, iv + 270 pages, 72 plates.

Kahn, John Ellison, editor
   1990   *Reader's Digest Illustrated Reverse Dictionary.* Reader's Digest
           Association, Pleasantville, New York, 608 pages, many
           unnumbered figures.

Kauffman, Henry J.
   1999   *Henry's Dutch Country Anthology, Volume II.* Masthof Press,
           Morgantown, Pennsylvania, [iv] + 90 pages, many
           illustrations.

Kinney, Thomas A.
   2004   *The Carriage Trade: Making Horse-Drawn Vehicles in America.*
           Johns Hopkins University Press, Baltimore, Maryland, xiv
           + 381 pages, 25 figures, 11 tables. (includes "essay on
           sources," pages 363–375, but no bibliography as such.)

Kugler, Georg
   1996   *Die Wagenburg in Schönbrunn.* Electa, Mailand Elemond
           Editori Associati, Kunsthistorisches Museum, Wien,
           Austria, 48 pages, many unnumbered illustrations.

Kunkel, Cindy
    1989   Berks County Heritage Center, Reading, Pennsylvania: the Eighth Annual Celebration. *Carriage Journal*, 27(3):141, 1 unnumbered figure.

Landau, Sidney I.
    2001   *Dictionaries: The Art and Craft of Lexicography*. Cambridge University Press, Cambridge, England. Second edition, xvi + 477 pages, 34 illustrations.

Lemon, James T.
    1972   *The Best Poor Man's Country: A Geographical Study of Early Southeastern Pennsylvania*. Johns Hopkins Press, Baltimore, Maryland; paperback edition, 1976, W. W. Norton, New York, xx + 295 pages, 59 figures, 34 tables.

Long, Amos, Jr.
    1972   *The Pennsylvania German Family Farm*. Breinigsville, Pennsylvania, The Pennsylvania German Society, Publications of the Pennsylvania German Society, Volume VI, 518 pages, many unnumbered illustrations.

Longstreet, Stephen
    1952   *A Century On Wheels: The Story of Studebaker: A History, 1852–1952*. Henry Holt , New York, reprinted 1970 by Greenwood Press, Westport, Connecticut, xiv + 121 pages, several unnumbered figures.

Mathews, M. M.
    1933   *A Survey of English Dictionaries*. Oxford University Press, Humphrey Milford, London, [iii] + 123 pages.

Mathews, Mitford M., editor
    1951   *A Dictionary of Americanisms, on Historical Principles*. University of Chicago Press, Chicago, Illinois, xvi + 1946 pages, many unnumbered figures.

McGaw, Judith A.
    1994   "So much depends upon a red wheelbarrow": agricultural tool ownership in the eighteenth-century mid-Atlantic. Pages 328–357, 1 figure, *in* Judith A. McGaw, editor, *Early American Technology: Making and Doing Things from the Colonial Era to 1850*. University of North Carolina Press, Chapel Hill, North Carolina (published for the Institute of Early American History and Culture, Williamsburg, Virginia).

Mercer, Henry C.
1975   *Ancient Carpenters' Tools, Together With Lumbermen's, Joiners' and Cabinet Makers' Tools in Use in the Eighteenth Century.* Fifth edition, for Bucks County Historical Society, Doylestown, Pennsylvania, by Horizon Press, xii + 339 pages, 250 figures. (First edition 1929; most recent edition by Dover Publications, Mineola, New York.)

Miles, William
1970   The good old Jackson wagon. *Carriage Journal,* 8(3):121–127, 8 unnumbered figures.

Miller, Lynn R.
2000   *Haying With Horses.* Small Farmer's Journal, Sisters, Oregon. 368 pages, many unnumbered illustrations (a mine of information about horse-drawn equipment, as are his several other books, and the Journal).

Mischka Press
1994–   *America's Rural Yesterday Appointment Calendars.* Mischka
2005   Press, Cedar Rapids, Iowa, 55 full-page photos in each, the majority by J. C. Allen (Mischka Farm, Oregon, Wisconsin, through 2004).

Morrison, Bruce, and Joyce Morrison
2003   *Wheelwrighting: A Modern Introduction.* Second edition, Cottonwood Press, DeWinton, Alberta, Canada, [viii] + iv + [ii] + 371 pages, 230 photos, 147 drawings.

Omwake, John
1930   *The Conestoga Six-Horse Bell Teams of Eastern Pennsylvania.* Ebbert & Richardson Co., Cincinnati, Ohio, 163 pages, frontispiece, many unnumbered illustrations. ("Published by John Omwake for private distribution.")

Parry, David
1979   *English Horse Drawn Vehicles.* Frederick Warne, London, 95 pages, 38 colored plates, 14 line drawings.

Pei, Mario A.
1970   Introduction. Pages v–vi, *in* Noah Webster, 1828, *An American Dictionary of the English Language,* volume I, reprint by Johnson Reprint Corporation, New York.

Peloubet, Don, editor
  1996   *Wheelmaking, Wooden Wheel Design and Construction.*
        Compiled by the Carriage Museum of America. Astragal
        Press, Mendham, New Jersey, 245 pages. Numerous
        illustrations. (Mostly reprints 19th century, and a few 20th
        century, writings on wheelmaking methods and
        machinery. Not definitive, for example, omits Archibald
        patent hubs, and inadequate on history of one-piece bent
        felloes.)

Pennsylvania Department of Labor and Industry
  1920   *Third Industrial Directory of Pennsylvania, 1919.* Department
        of Labor and Industry, Harrisburg, Pennsylvania, 1212
        pages.

Piggott, Stuart
  1992   *Wagon, Chariot and Carriage: Symbol and Status in the History
        of Transport.* Thames and Hudson, New York, 184 pages,
        33 illustrations.

Posey, John Thornton
  1989   The improvident ferryman of Mount Vernon: the trials of
        Captain John Posey. *Virginia Cavalcade,* 39(1):36–47, 9
        unnumbered illustrations.

Reber, Ronnie Lewis
  1992   *Reber Wagon Works, Centreport, Pennsylvania, Established
        1892. Drift Anthracite Coal Company, Bowmanstown,
        Pennsylvania, 1912 thru 1950.* Self published, 47 pages, many
        unnumbered illustrations. (Year of publication not
        indicated in the book, but confirmed by author, personal
        communication, 2004; pages 1–31 on wagons. Very rare,
        cataloged in Library of Congress under the number 93-
        112785.)

Reist, Arthur L.
  1975   *Conestoga Wagon – Masterpiece of the Blacksmith.* Self
        published, ii + 50 pages, many unnumbered illustrations.
        (Very rare, cataloged in Library of Congress under the
        number 75-34956.)

Salaman, R. A.
  1990   *Dictionary of Woodworking Tools C. 1700–1970, and Tools of
          Allied Trades.* Revised edition, Taunton Press, Inc.,
          Newtown, Connecticut, 546 pages, 741 figures (many
          consisting of multiple images). (First published in Great
          Britain 1975; revised British edition, 1989.)

Sellens, Alvin
  2002   *Dictionary of American Hand Tools: A Pictorial Synopsis.*
          Schiffer Publishing Ltd., Atglen, Pennsylvania, xii + 546
          pages, numerous illustrations.

Seymour, John
  1990   *The Forgotten Crafts.* Portland House, Outlet Book
          Company, Random House, New York, 192 pages,
          numerous unnumbered illustrations (first published 1984
          by Dorling Kindersley, London).

Shank, Michael W.
  1972   Rural Lancaster County technology of the 1850's.
          Community Historians Annual, 11(6):IV + 37 pages, 11
          figures. (Revised from Shank's 1969 thesis for the Master
          of Education degree at Millersville State College.)

Shumway, George, and Howard C. Frey
  1968   *Conestoga Wagon 1750–1850; Freight Carrier for 100 Years of
          America's Westward Expansion.* Third edition, George
          Shumway, York, Pennsylvania, XII + 281 pages, 131 figures.

Smith, D. J. M.
  1988   *A Dictionary of Horse-Drawn Vehicles.* J. A. Allen & Co. Ltd.,
          London, [viii] + 189 pages.

Smith, Peter Haddon
  1971   *The Industrial Archeology of the Wood Wheel Industry in
          America. A Dissertation Submitted to the Faculty of the Graduate
          School of Arts and Sciences of The George Washington
          University in Partial Satisfaction of the Requirements for the
          Degree of Doctor of Philosophy.* xii + 312 pages. Many
          illustrations. (Dissertation Abstracts, 1972, Volume 33, No.
          1, Series A, pages 261A-262A, Order no. 72-9014.) (Based
          predominantly on three Pennsylvanian resources: Hoopes,
          Bro. & Darlington, Wheelmakers of West Chester; Bareville
          Woodcraft Company, Wheelmakers of Leola; and Kube's
          study of the Gruber Wagon Works, Mt. Pleasant.)

Spivey, Towana, editor
    1979   A historical guide to wagon hardware & blacksmith supplies. *Contributions of the Museum of the Great Plains, Lawton, Oklahoma,* no. 9, iii + 202 pages, many unnumbered illustrations.

Stratton, Ezra M.
    1878   *The World on Wheels; Or, Carriages, With Their Historical Associations from the Earliest to the Present Time, Including a Selection from the American Centennial Exhibition.* Published by the author, New York, [vi] + 489 pages, frontispiece, nearly 400 unnumbered illustrations.

Strohm, John
    1863   The Conestoga horse. *Report of the Commissioner of Agriculture for the Year 1863,* Washington, Government Printing Office, pages 175–180, plate XXIV.

Sturt, George ("George Bourne")
    1923   *The Wheelwright's Shop.* The University Press, Cambridge, xii + 236 pages, 8 plates, 24 text-figures. (Reprinted repeatedly, including paperback as recently as 1993, which, n.b., does not include the plates, but does include a foreword, pages viii–xiv, by E. P. Thompson, dated 1992.)

Telleen, Lynn
    2001   [Hansen Wheel & Wagon Company, Letcher, S.D., est. 1978.] *Draft Horse Journal,* summer 2001. (Article included on company website, without full citation to Journal, whose office refused to provide citation in response to telephone request of 24 November 2004.)

Thompson, John
    1980   Horse-drawn farm vehicles. *Sourcebooks* No. 8, John Thompson, Fieldway, Fleet, Hampshire, United Kingdom, 78 + [2] pages, many unnumbered illustrations.

    1983   The wheelwright's trade. *Sourcebooks* No. 10, John Thompson, Fieldway, Fleet, Hampshire, United Kingdom, 70 + [2] pages, many unnumbered illustrations.

Unger, Harlow Giles
    1998   *Noah Webster: The Life and Times of an American Patriot.* John Wiley & Sons, New York, xiii + 386 pages, frontispiece, 21 unnumbered illustrations.

United States Bureau of the Census, Department of Commerce
> 1924   *Biennial Census of Manufactures, 1921. Group 12. — Vehicles for Land Transportation.* Government Printing Office, Washington, pages 1007–1022, tables 845–857. (Includes some retrospective data back to 1859; similar, but shorter, publications for 1931, 1933, and 1937.)

Vince, John
> 1975   *An Illustrated History of Carts and Wagons.* Spurbooks, Bourne End, Buckinghamshire, United Kingdom, 160 pages, frontispiece, many unnumbered illustrations.

> 1987   *Discovering Carts and Wagons.* Third edition, Shire Publications, Aylesbury, United Kingdom, 72 pages, 32 photographs and many unnumbered drawings.

Vineyard, Ron
> 1993   *Virginia Freight Waggons, 1750–1850.* Unpublished document of Colonial Williamsburg Foundation, Williamsburg, Virginia, iv + 191 pages, 36 figures. (Unpublished report on Virginia variant of Conestoga wagon.)

Wendel, C. H.
> 1981   *150 Years of International Harvester.* Crestline Publishing Company, Sarasota, Florida, 416 pages, many unnumbered illustrations.

> 1991   *150 Years of J. I. Case.* Crestline Publishing Company, Sarasota, Florida, 336 pages, many unnumbered illustrations.

> 2004   *Encyclopedia of American Farm Implements & Antiques.* Second edition, Krause Publications, Iola, Wisconsin, 496 pages, more than 2500 illustrations.

Wheeling, Ken
> 1991   From wheelbarrows to wagons: The story of the Studebaker Brothers Manufacturing Company. *Carriage Journal,* 29(1):21–24, 5 unnumbered figures.

> 1992A   Trans-Mississippi transport: Part IV; Wheeled transport: The wagons that went west. *Carriage Journal,* 30(1):29–32, 7 unnumbered figures.

> 1992B   Trans-Mississippi transport: Part V; Louis Espenschied and his freight wagons. *Carriage Journal,* 30(2):66–69, 6 unnumbered figures.

> 1992C   Trans-Mississippi transport: Part VI; Joseph Murphy and the "Murphy wagons." *Carriage Journal,* 30(3):115–119, 5 unnumbered figures.

1993 Trans-Mississippi transport: Part VII; The Schuttler wagons: Triumph and tragedy. *Carriage Journal*, 30(4):154–158, 8 unnumbered figures.

1994A The Bain Wagon Company. *Carriage Journal*, 31(4):166–171, 12 unnumbered figures.

1994B The American Waggon Association. *Carriage Journal*, 31(4):179–180, 3 unnumbered figures.

1995 The Thornhill Wagon Company. *Carriage Journal*, 33(2):55–58, 8 unnumbered figures.

1996A The Leudinghaus side. *Carriage Journal*, 34(2):78–79, 4 unnumbered figures.

1996B The D. W. Scobie Wagon Factory. *Carriage Journal*, 34(3):115, 1 unnumbered figure.

1997 The Mitchell-Lewis Wagon Company. *Carriage Journal*, 34(4):156–160, 8 unnumbered figures.

2000 Hansen Wheel and Wagon Shop. *Carriage Journal*, 35(2):70–73, 10 unnumbered figures.

Wigginton, Eliot, editor

1973 *Foxfire 2*. Anchor Press/Doubleday, Garden City, New York, 410 pages, frontispiece, 387 plates. (Pages 118–141, plates 79–125, on wheel- and wagon-making.)

Wigginton, Eliot, and Margie Bennett, editors

1986 *Foxfire 9*. Anchor Press/Doubleday, Garden City, New York, xvi + 493 + [3] pages, frontispiece, 482 plates. (Pages 267–320, frontispiece, plates 183–346, on wheel- and wagon-making.)

NOTE ADDED IN PROOF: Consistent with my assertion on page 18 that the coverage of wagon-making in the British Isles is better than that in the United States is yet another book that I neglected to cite, but which deserves to be more widely known; to wit:

Quennell, Marjorie, and C. H. B. Quennell

1961 *A History of Everyday Things in England. Volume III, 1733 to 1851.* B. T. Batsford, London, United Kingdom, 220 pages, 170 figures. (Sixth edition cited here; see also other editions and other volumes.)

Although ostensibly written for children and covering a broad range of topics, the writers used authoritative sources on all subjects and clearly had a special interest in wagons, as they did original study and drawings themselves (pages 132–142, figures 100–109); and campaigned in 1919–1920 for preservation of wagons. Let us hope that British school children continue to be so fortunate!

Clayton E. Ray

Acknowledgments

Foremost, we and all who care about wagons owe thanks to the men of the Gruber family, who remained true to their trade for half a century after it had generally succumbed to the power of petroleum; and to the men and women of the Gruber family for their dedication to the preservation of the Wagon Works and the promulgation of its history. Similarly, we are indebted to all of the farmers of southeastern Pennsylvania, who have continued horse-powered farming successfully in modern times. Specifically for this project, Paul Kube applied his special interests and attributes at the right moment to capture history while it lived. To him and to all of the individuals, organizations, and governmental entities who worked long and hard to make the move and preservation of the Wagon Works a reality, we owe lasting gratitude.

For the work on this book, the approval and assistance of the Kube family was indispensable. These include Paul Kube's widow, Sophie (her son, Joe Dugan, and niece, Antoinette Haas, neither related to Paul Kube), Paul Kube's daughter, Sallie Fisher (and her husband, William Fisher), along with Kube's niece, Carol Miller.

The greatest fringe benefit to one's interest in the horse-drawn world is the opportunity it affords to interact with people who are themselves interested, interesting, and generous. By no means listed in order of importance, but in part grouped by affinity, we thank: Susan Green, Carriage Museum of America, and David J. Bohaska, Claire Catron, James Roan, and Roger White, Smithsonian Institution, all for repeated help with literature; Ronnie Reber, Art Reist, and Charles Schrack for sharing their knowledge of wagons; Major General (Retired) James A. Johnson, Martin H. Johnson, and Carol L. Wahlin for information about wagon-making in Stoughton, Wisconsin; Ann Moore and Robert Murray for information about the Foxfire Museum and its wagons; Deborah Walk and an anonymous volunteer for information on the Ringling and other circus museums; Jane S. Bray, Cheryl T. Desmond, and Marilyn Parrish, of Millersville University, for affirmative response to our query regarding the propriety of our publishing Paul Kube's Millersville thesis; Bill Burgess, Don Cornett, Doug Hansen, and Harlin Olson, for information about their operational West tire setters and other aspects of their wagon work; Richard O'Connor, National Park Service, for information from their files on the Wagon Works; Charles W. Cook, of R. S. Cook and Associates, Inc., general contractors for the move of the Wagon Works in 1976, for providing information on the move and for sharing his video; Rob Deisemann, volunteer, and Rob Bierbower and Daniel Roe, seasonal staff, at the Berks County Heritage Center, for help with Gruber production records; Sue Eddinger, secretary at the Berks County Heritage Center, for assistance with numerous and diverse tasks; Cyndi Nasta, Joel Palmer, Glenn Riegel, and Jonathan L. Shalter, III, for photos; Bruce and Carol Hunsberger, for keeping us informed about their work and for sharing their knowledge. Susan Green also read the manuscript and gave us the benefit of her wide and deep knowledge of the horse-drawn world.

Virtually without exception, all those with special knowledge and/or position have been extraordinarily responsive to our calls for help.

Finally, the Central Rappahannock Regional Library, Fredericksburg, Virginia, deserves our best thanks for uniformly good service in finding literature and providing computer facilities for communication and preparation of manuscript.

Index

ABOUT THE AUTHORS: **Paul A. Kube** (1918–1988) served in the US Navy from 1937 to 1960, then attended Millersville State College (now University), Millersville, Pennsylvania, where he earned Bachelors (1963) and Masters (1968) degrees in Education. He taught at Boyertown Area High School from 1963 until 1971. **Clayton E. Ray,** Ph. D. (Harvard, 1962), Curator Emeritus in the Department of Paleobiology at the National Museum of Natural History, Smithsonian Institution, presently lives on a farm near Fredericksburg, Virginia, where, among other activities, he collects and uses horse-drawn farm equipment. **Cathy L. Wegener,** B.S. (Pennsylvania State University, 1981), is Superintendent of Interpretive Services for the Berks County Parks & Recreation Department, in which position she manages the operation, preservation, and interpretation of the Gruber Wagon Works.